1950•1975
FARM TRACTORS

Lester Larsen

American Society of
Agricultural Engineers

FIRST EDITION

©1981 by the American Society of Agricultural Engineers. All rights reserved. No part of this book may be used or reproduced in any manner without the express written consent of the publishers except in the case of brief quotations embodied in critical reviews or articles about the book. For information address the publishers, the American Society of Agricultural Engineers.

LCCN: 81-69655
ISBN: 0-915150-36-4

Printed in the United States of America by the American Society of Agricultural Engineers, 2950 Niles Road, P.O. Box 410, St. Joseph, Michigan 49085.

Table of Contents

About the Author • v
Carlton L. Zink • vi
Acknowledgement and Photo Credits • vii
Reference Publications • viii
Introduction • 1
1950 • 7
1951 • 13
1952 • 17
1953 • 23
1954 • 29
1955 • 33
1956 • 37
1957 • 45
1958 • 51
1959 • 59
1960 • 69
1961 • 79
1962 • 83
1963 • 87
1964 • 93
1965 • 97
1966 • 107
1967 • 111
1968 • 117
1969 • 123
1970 • 131
1971 • 137
1972 • 145
1973 • 153
1974 • 159
1975 • 165
Appendix: Tractors Tested at Nebraska
1950-1975 • 178

DEDICATION

This book is dedicated to the agricultural engineers and others who contributed toward the development of the modern tractor and to those who cooperated by providing needed information and photographs to complete this book.

ABOUT THE AUTHOR

Few people are as well qualified as Lester Larsen to compile a work as extensive as this one. For nearly 30 years Larsen was the engineer-in-charge of the Nebraska Tractor Testing Laboratory. Begun in 1919, the program was designed to ensure that tractors bought and sold in the state were in good working order. Under Larsen's leadership, beginning in 1946, the program expanded to encompass more performance measurements, fuel consumption, noise, and other factors which were continually updated. The Laboratory gained such a reputation for reliability and usefulness that the annual reports became the "bible" for tractor dealers and manufacturers. They are still much in demand the world over.

Larsen brought much more than technical expertise to the program. His integrity in administering the program is known and admired. His rapport with the farmer made him a sought-after speaker. His dedication to agriculture and belief in the benefits of the testing program moved him to help leaders in other countries establish similar programs.

Holding two degrees from the University of Nebraska, Larsen was Professor Emeritus there when he retired in 1975. He has authored many extension service bulletins, technical papers, magazine articles, and handbooks. An active member of the American Society of Agricultural Engineers for many years, he was awarded the Cyrus Hall McCormick Gold Medal in 1976 for "exceptional and meritorious engineering achievement in agriculture."

Larsen's only regret in a long career of achievement may be that his dream of a great tractor museum at the University of Nebraska never came true. There are a few old-timers on display there—Fordson, an old John Deere, a Farmall, the first Ferguson—but not the comprehensive museum Larsen always wanted. This encyclopedic listing, as complete as Larsen could make it, may well exceed that original dream. It could actually be thought of as a museum-without-walls, easily within reach of many. It most certainly should be thought of as a lasting tribute to Larsen's expertise and his dedication to the world of mechanized farming.

Samuel H. Rosenberg
ASAE

The author wishes to extend special thanks to Carlton L. Zink, Engineer-in-Charge of the Nebraska Tractor Tests from 1930 to 1941. Without his diligent help and the benefit of his expertise, this book may not have been possible.

Though now retired, Mr. Zink still works as a safety consultant, writing and lecturing on matters related to Agricultural Product Safety. He has worked both in industry and in the public service. He is active on committees of the Society of Automotive Engineers, American Society of Agricultural Engineers, Farm Conference of National Safety Council and the National Institute for Farm Safety. His contributions to agricultural engineering, throughout his life and in this book, are worthy of heartfelt appreciation.

ACKNOWLEDGEMENT AND PHOTO CREDITS

Over five years of research and travel went into the preparation of **Farm Tractors: 1950-1975**. Many individuals were helpful and considerate in providing photos and information for this publication. The author particularly acknowledges the following:

Sam Rosenberg, ASAE Headquarters Staff, who as the author's editor for the past five years, was encouraging from first draft to last. Sam revised and edited **The Agricultural Tractor: 1855-1950**, originally prepared by the late R. B. Gray. With his constant prodding, advice, friendship and purpose this book became reality.

Special recognition is due **Frank Walters** who has given freely of his time and provided valuable advice in preparing the draft. He also helped find resources for information needed in making this publication possible. Frank is a staff engineer at John Deere Product Engineering Center, Waterloo, Iowa. He is a long time member of ASAE and SAE technical committees; and chairman of the U.S. Technical Advisory Group for International Standards Committee, ISO/TC 23, Tractors and Machinery for Agriculture and Forestry.

Allis-Chalmers: W. R. Bryant and R. Becker, C. Cannon, J. R. Hoepfl, R. Johansen, W. E. Muelhausen, M. Pribyl

Belarus Machinery: O. Kroulkevitch and M. W. Kach, A. D. Kryzhanovsky, F. G. Rollins

Borg-Warner, Marvel-Schebler/Tillotson Division: G. D. LaMasters

Deutz Tractor Corporation KHD: F. Kaplan and R. Wesner

D. L. Curtis (Rite Tractor): D. Curtis and D. L. Dillman

Deere and Company: F. Walters and R. Candee, T. Corners, M. Miller, V. J. Schultz, F. Zoz

Donaldson Company: R. E. Larson and G. H. Ilika

Behlen Manufacturing Company: W. Behlen and R. Ferris

Ford Tractor Operations: D. J. Campbell and R. Dewey, J. B. Jackson, J. Zich

Garrett Airesearch: P. M. Uitte

Four-Wheel-Drive Power Horse: S. Bonham and H. Witzel

Hefty Tractor Company: R. H. Ziorgen

Implement and Tractor: J. Root

Intercontinental Manufacturing Company: R. Shape and M. Thiemann

International Harvester Company: C. Stropes and R. Coleman, G. Lennes, R. E. Reed, H. J. Slothower

J. I. Case (David Brown): E. A. Heys and D. Brown, R. Dobler, W. O. Gipp, K. E. Hohnl, M. Ladwig, J. E. Langdon, I. M. Robson

Knudson Tractor: J. Knudson and R. Allmand

Kubota Tractor Ltd.: H. Hanaoka and Y. Horii, O. Yamasaki

Leyland Tractors (Nuffield): C. M. Chance and R. Bassett, N. Newton

Long Tractors: M. Saunders and J. Long, W. Long

Massey-Ferguson: B. Wardlow and E. Barns, R. Cleaves, R. M. Doll, J. Fox, H. Lange, D. Manson, E. Zeglen

M & W Gear Company: E. Minor and M. Cornwell

M-R-S Tractor: D. White and W. Moree, J. Nelson

Northern Manufacturing Company (Big Bud Tractor): R. Harmon and D. Zavas

Rome Tractor (Woods and Copeland): S. Caughran and J. Copeland

Satoh Agricultural Machinery Ltd. (Mitsubishi Agricultural Machinery Ltd.): M. Ono and S. Shoji, P. Stankee, J. Takeda

Slope Tractor: G. Tregallis and R. Diehl, O. C. Richards

Stanadyne Diesel Systems: K. Davis and J. Boissonneault

Steiger Tractor: L. Moen and S. Anderson, S. Melroe, J. Mortensen, B. Quigley, C. Zick

Versatile Manufacturing Ltd.: M. Klein and W. Lenius, J. Lutz, E. Ukraince

Wagner Tractor: W. E. Wagner and J. Hauenstein, W. E. Larsen

White Farm Equipment (Oliver Corp., M-M & Cockshutt): G. Lockie and H. Dommel, M. Dunham, J. Janssen, A. J. Meyers, C. Overlee, R. Prunty, C. Reller, D. Starr, M. Walker, L. Zetoka, J. Zellars

Yanmar Tractors: T. Watson and H. Ura

Most of the information and photos were obtained through the cooperation of the various tractor manufacturers and their representatives. Also, information, data and photos were obtained from the files of the Nebraska Tractor Testing Laboratory. This was possible through the cooperation of Lou Leviticus, Mike Mumgaard, John Carlile, Dave Morgan, Brent Sampson and Ray Beckner.

From the universities and other institutions: C. B. Richey, L. W. Knapp, Kenneth Domier, W. F. Buchele, George Larson, W. J. Promersberger, W. E. Larson, William Splinter, R. W. Kleis, George Steinbruegge, Rollin Schnieder, Kathleen Cifra (F. Hal Higgens Library), University of Nebraska Department of Information and C. Y. Thompson Library.

From others interested in tractor development: Carlton Zink, T. H. Morrell, Walter Behlen, Elmo Minor, Merlin Hansen, Harold Moberg, Ted Worrall, Ned Meier and Leroy Barry.

Much care has been taken to prepare **Farm Tractors: 1950-1975** to insure accuracy.

The author will welcome comments and suggestions that might be used in future editions.

REFERENCE PUBLICATIONS

A History of Progress in Mechanization, (ASAE) 1976, W. J. Promersberger

A Global Corporation, E. P. Neufeld

Agricultural Engineers Yearbook (ASAE)

An Industrial Heritage, W. F. Peterson

An Historical Perspective of Farm Machinery, (SAE) 1980

A Free-Piston Turbine Power for Farm Use, (SAE) 1957, R. L. Erwin

A Free-Piston Tractor Power Plant, (SAE) 1957, O. B. Norden

Experiences in North Dakota with Safety Frames on Highway Tractors, (ASAE) W. I. Hansen

Encyclopedia of American Farm Tractors, C. H. Wendel

Farm Tractors in Colour, M. Williams

Harry Ferguson, C. Fraser

Highlights of Ford in Agriculture, Ford Tractor Division, Ford Motor Co.

John Deere Tractors 1918 to 1976, Deere and Company

Liquified Petroleum Gas for Tractors, Kansas Experiment Station #71, G. Larson

Machines of Plenty, S. G. Charlton, J. I. Case Co.

Men, Machines and Land, FIEI, 1974

Milestones in the Application of Power to Agricultural Machines, (SAE) 1980, N. Coleman and K. W. Burnham

Noise Reduction, M. L. Mumgaard, Nebraska Tractor Testing Laboratory

Noise Sound without Value, R. D. Schnieder, Univ. of Nebraska

Official Guide-Tractors and Farm Equipment, National Farm & Power Equipment Dealers Association

Progress in Tractor Power From 1898, White Motor Corporation

Safe Tractor Operations, Nebraska Extension Service, R. D. Schnieder

Safety Accomplishments on Agricultural Equipment, (SAE) 1980, C. Zink

Sound Level Tests of Tractors at The Nebraska Tractor Test Laboratory, 1971, W. E. Splinter

Some Observations on Noise and the Testing of Tractors at the University of Nebraska, 1971, G. W. Steinbruegge

The Nebraska Tractor Test Data, 1950 to 1975, Univ. of Nebraska

The Development of Agricultural Equipment Power Take-Off Mechanism, (SAE) 1980, T. H. Morrell

The Effect of Tractor Noise on the Auditory Sensitivity of Tractor Operators, 1958, Univ. of Iowa, Lierle, Scott and Reger

Tandem Tractors for More Tillage Power, (ASAE) 1962, C. B. Richey

Thirty Years of Nebraska Tractor Testing, ASAE 1976, L. F. Larsen and Lou Leviticus

Tractors and Their Power Units, Barger, Liljedahl, Carlton and McKibben

Tractor Cab Noise Control, (ASAE) J. Egging, G. Schmer and K. McMillen

Tractor Safety Cabs, Test Methods and Experience Gained During Ordinary Farm Work in Sweden, (1954) H. A. Moberg

The Agricultural Tractor: 1855-1950, (ASAE) R. B. Gray

Tractors Produce Ear Damaging Noise, (1965) Univ. of Nebraska, Simpson and Deshazer

Turbochargers, H. MacInnes

Understanding the New Noise Tests at Nebraska, (1971) "Implement & Tractor", C. Floyd

Introduction

INTRODUCTION

This book is intended to be a sequel to a previous publication by the American Society of Agricultural Engineers, The Agricultural Tractor: 1855-1950, the book prepared by the late R. B. Gray, who had been in charge of the Farm Machinery Section, Agricultural Research Service, USDA. This book follows basically the same pattern as Gray's book. Most of the tractors sold in the United States from 1950 to 1975 are included, except crawler and garden tractors.

Numerous references to tractor testing at the University of Nebraska's Tractor Testing Laboratory at Lincoln, will appear in this book, because of the author's close association with the laboratory over many years. These experiences are reflected in discussions of the various phases and changes that have occurred in tractor development over a 25-year period. This was a period when many major changes took place, not only in the horsepower output but also in tractor safety, operator comfort, and ease of operation.

Tractor Power Increases

Prior to 1950, the horsepower of agricultural tractors was comparatively small by present-day standards. In 1950, few wheel-type tractors exceeded 50 maximum belt horsepower, and most farm tractors were considerably below that size. But by that time, it became apparent that tractors were increasing in horsepower—the horsepower race was starting. However, engineers designing tractors then did not dream that horsepower would reach the levels which today's tractors have achieved.

The Horsepower Confusion

Most of the tractors listed in this book were tested at the Nebraska Tractor Testing Laboratory. Most of the horsepower ratings mentioned were taken from results obtained during official laboratory tests.

The Tests were conducted according to the Agricultural Tractor Test Code approved by the American Society of Agricultural Engineers (ASAE) and the Society of Automotive Engineers (SAE).

The maximum horsepower, belt or PTO results were taken during observed conditions and none include any correction factors. The tractors that were not tested at Nebraska may show similar advertised power, but probably do not. Because of a number of terms relating to horsepower now in use, the reader may wish to refer to the meaning of various terms presented in a 1965 ASAE technical paper by H. J. Slothower, Product Engineer, International Harvester Co. His definitions and explanations are:

- **Indicated horsepower** is that developed within the cylinders of the engine. This will be high. These results are not often quoted in advertisements.
- **Gross horsepower** is the power measured at the flywheel of a bare engine or stripped engine not installed in a tractor. In this case the engine is equipped with only the parts needed to make the engine run.
- **Net horsepower** is the power obtained at the flywheel while the engine is equipped with all the accessories needed to make it operate by itself.
- **PTO horsepower** is the measured power obtained at the tractor power take-off or PTO. The PTO or belt tests made during the Nebraska tractor tests do not reflect any corrected results but are only observed results.
- **Drawbar horsepower** is the power available at the drawbar of the tractor to pull equipment. The Nebraska tractor drawbar tests are only observed results; no corrections are made for temperature, humidity, barometric pressure or wheel slippage.
- **Peak brake horsepower** is the highest horsepower developed without any speed limitation. These runs often are made during a very short time.
- **Maximum intermittent brake horsepower** is considered the maximum saleable horsepower and usually is used when referring to power units.
- **Continuous-brake horsepower** ratings are those recommended for operation under continuous conditions. This is used mainly for power unit terminology.
- **SAE horsepower** formula is intended to determine the approximate brake horsepower of an engine.

$$\text{Horsepower} = \frac{D^2 \times N}{2.5}$$
$$\left[\begin{array}{l} D = \text{the bore of the cylinder in inches.} \\ N = \text{the number of cylinders.} \end{array} \right]$$

The SAE (Society of Automotive Engineers) formula is a simplification of the indicated horsepower. It assumes that the mean effective pressure is 90 pounds per square inch, that the piston speed is 1000 feet per minute and that the mechanical efficiency is 75 percent.

To add to the confusion of horsepower terms, correction factors also are used to make the results of engines more nearly comparable when operated under various atmospheric conditions. The full amount of corrected horsepower will seldom be available for useful work. Correction factors include corrections for altitude, temperature and humidity. This makes it difficult to know which advertised horsepower figure is reasonable unless it is specified completely.

The pressed steel pan seat was standard equipment on most farm tractors of the 1920's and 1930's.

Operator Comfort Considered

In the earlier years, manufacturers of farm tractors gave little thought to the operator's comfort. Pressed steel pan seats, now considered collectors' items, were used on tractors for many years. If a farmer placed a pad on the seat or raised an umbrella to protect himself from the sun, he generally was viewed as a sissy. Many operators often preferred to stand while driving rather than sit on the hard seat. Standing was not the most comfortable position either, because often there was no convenient place to stand.

Various cabs had been used for many years, even on some of the old steam traction engines. Many of these cabs had only a cover over the top. In 1938, Minneapolis-Moline introduced the "COMFORTRACTOR" which had an enclosed cab, making the tractor resemble a

Minneapolis-Moline introduced its Comfortractor in 1938. The cab was designed mainly to protect the operator from snow, rain and cold weather. Safety glass and a cushioned seat for two were other selling points. (Implement & Tractor)

This photo was taken of Senator Arthur Capper, during Minneapolis-Moline "Comfortractor" Days in Topeka, Kansas, Dec. 5, 1938. (Implement & Tractor)

streamlined sport roadster. The cab was intended mainly to protect the operator from cold weather, snow, and rain. Little thought was given to roll-over protection, or the elimination of loud engine and gear noises.

Safety Features Come Slowly

As early as 1947, several safety devices were tried in an effort to help prevent tractor accidents. Most notable was the safety switch that would shut off the ignition if the operator left the tractor seat. Mercury switches also were used to shut off the ignition when tractors reached a certain degree of slope, but in case of an over-turn, this generally was too late to save the driver. Most farmers didn't want to be bothered with such devices, which they termed "foolishness," and usually removed them after a short trial period.

The 4-H tractor club program, sponsored by the Standard Oil Company of Indiana in 1944, helped stimulate tractor safety. This program encouraged boys and girls to maintain tractors in better condition and also to promote safety. Extension agricultural engineers developed model electric-drive tractors that could simulate a full-sized tractor in operation and show how tractors could over-turn when operated improperly. These model tractors were used in many demonstrations throughout the country.

Combustion Chamber Deposits

Following WW II, many farmers using gasoline in their tractors experienced valve failures. This problem was apparent to engineers. The L-head engines seemed

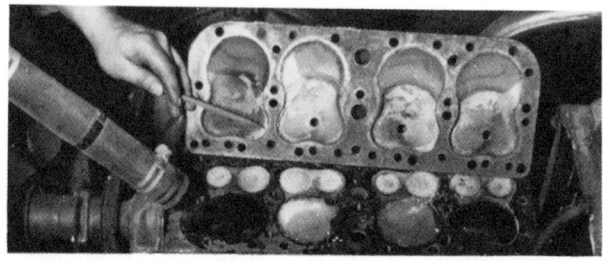

Following WW II combustion chamber deposits caused loss of power and frequent valve failures in tractor engines. Starting with a clean engine, it was not uncommon to lose 10 percent of the power after 10 or 12 hours of operation.

to be more susceptible than the I-head or valve-in-head engines. It was not uncommon to lose 10 percent of the tractor engine power after 10 or 12 hours of operation. Test reports often contained statements relating to this situation, such as: "During Test 'B', a decrease in horsepower occurred. The engine head was removed, combustion chamber cleaned, and the valves lapped lightly. The test was resumed with improved performance." These experiences no doubt influenced the change from L-head engines to the more popular I, or valve-in-head engines and valve rotators.

Fuels

When tractor testing began in 1920 two main fuels were available—kerosene and gasoline. By 1925 distillate was accepted as a replacement for kerosene. Actually, it was considered a better fuel. However, kerosene continued in use until 1934. The last tractor tested using kerosene was a McCormick-Deering W-12 (Test 229).

Distillate remained a popular and economical fuel for many years and continued in use until 1956. The last tractor tested using distillate was a John Deere 720 All-Fuel tractor (Test 606).

Gasoline proved the most desirable fuel following WW II. Gasoline engines started easier than those using distillate and dilution in the crankcase oil was less. Farmers liked using gasoline exclusively because then they needed to stock only one fuel on the farm, and the price of gasoline was acceptable at that time. Also, the gasoline engines with suitable compression ratios

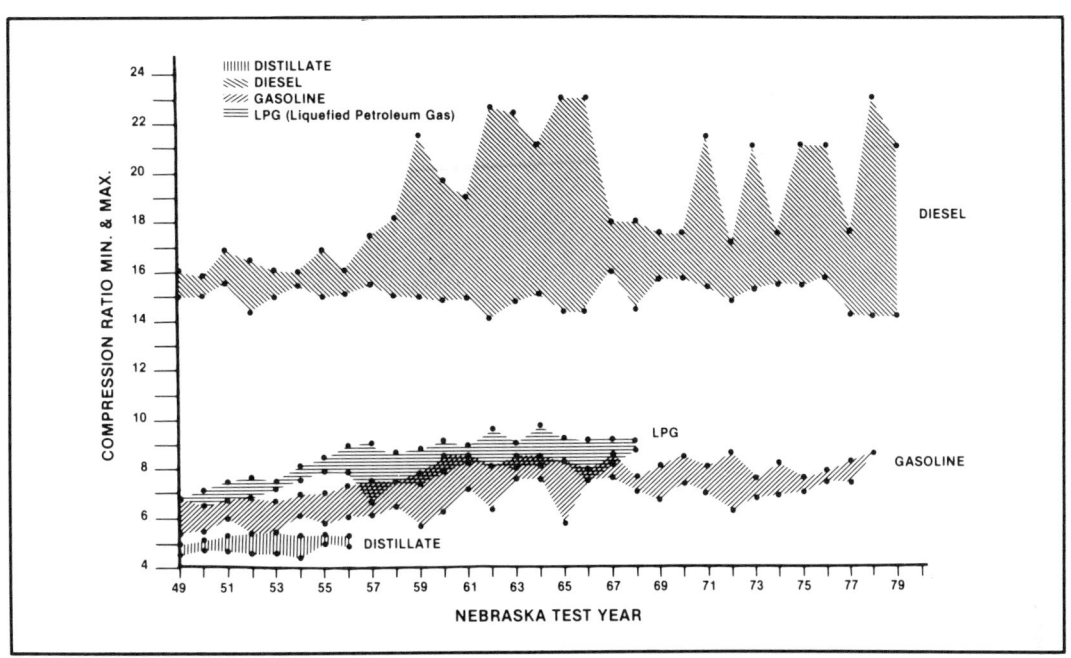

This chart prepared from Nebraska Tractor Testing Laboratory results shows compression ratios of four tractor fuels—distillate, diesel, gasoline, & LPG.

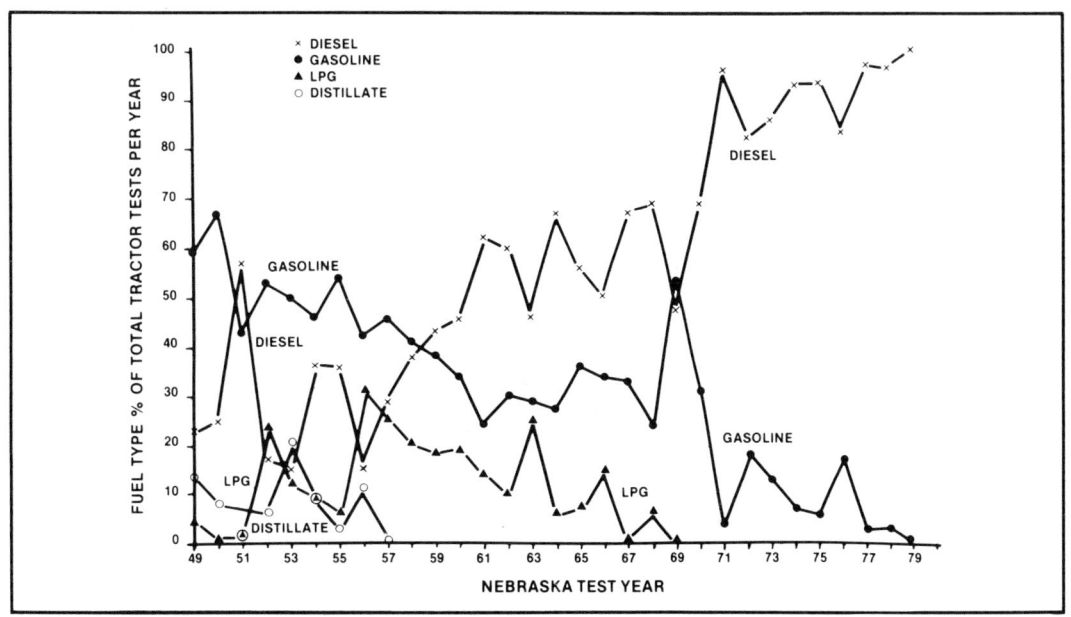

Note increase in diesel fuel and decrease in other fuels used by tractors tested at the Nebraska Tractor Testing Laboratory through the years.

1950-1975 FARM TRACTORS

developed more power for the same cubic inch displacement. Distillate-burning engines might have only a 4.7 to 1 compression ratio, compared to gasoline engines with a 7 to 1 compression ratio or higher.

The first diesel tractor tested (Test 208) at Nebraska was a Caterpillar. The year was 1932. In 1935, a McCormick-Deering WD-40 became the first wheel-type diesel tractor tested (Test 246). In 1949, the Laboratory tested 10 diesel tractors, indicating a trend toward the use of diesel fuel—a trend that was to continue for many years. The same year the Laboratory tested its first tractor using LPG or propane. It was a Minneapolis-Moline Universal Standard (Test 411).

LPG or propane (Butane in the South) proved to be an economical fuel, but its use slipped considerably after 1963 as diesel became the dominant fuel.

Power Take-Offs Introduced

The power take-off (PTO) for tractors appeared toward the end of WW I. International Harvester introduced a kerosene tractor with a PTO as optional equipment in 1918. This was believed to be the first American tractor to offer a practical rear PTO mechanism for propelling trailed implements.

The McCormick-Deering 15-30 was the first tractor tested (Test 87) which was equipped with PTO as standard equipment. The test was conducted in 1922. Oliver introduced *independent* PTO on its tractors in 1946. A year later the Nebraska Laboratory belt tested its first tractor equipped with an *independent* PTO (Test 382), a Cockshutt 30 gasoline tractor. In the years to follow, the *independent* PTO was to become a popular addition to the modern tractor.

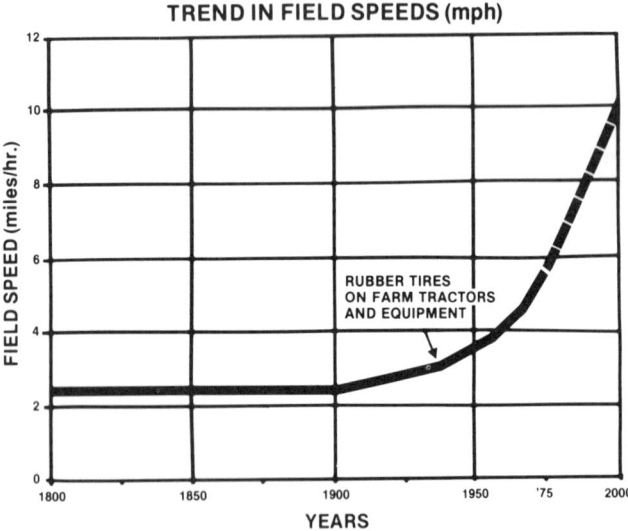

Field speeds of tractors increased dramatically after the tractors were equipped with rubber tires. (Courtesy W. J. Promersberger)

Travel Speeds Increase

The early tractor was intended to replace the horse, therefore the working speed of most steel-wheeled tractors was 2 to 2½ mph, or about the same speed as horses walked when working. This speed changed little until the mid-1930's when pneumatic rubber tires were added to farm tractors. Since then, working speeds have more than doubled. While transport speeds have increased to the upper teens. Many authorities consider this to be the limit for safe handling of a tractor.

Many older tractors in use during the mid-1940's were geared for a top speed of about 4 to 5 mph. This was about the maximum speed designed for steel-wheel tractors. However, by this time most farmers had changed their older tractors to pneumatic tires and were interested in increasing the road speed of these tractors. Several manufacturers provided auxiliary gear boxes to be installed on conventional slow-speed tractors. One of these manufacturers was the Behlen Company, Columbus, Nebraska. Behlen "bolt on" units were available for the F-20 Farmall, John Deere D and other tractors of that era. The Behlen road gear was designed to step up travel speeds to 14 mph. Among other manufacturers was M&W of Anchor, Illinois.

The Behlen road gear was a bolt-on device designed to step up travel speeds to 14 mph. (Courtesy Walt Behlen Universe)

Dry-Type Air Cleaners Appear

Nearly all tractors built during the past 20 years or more were equipped with oil bath air cleaners. The oil bath air cleaner was quite satisfactory if properly maintained, but it caused service problems for the operator. When neglected, which occurred often, it could cause severe engine failures. This problem prompted development work which eventually led to the dry-type air cleaner.

Foreign Tractors Tested

Prior to 1946, only one foreign tractor company had applied to have its tractors tested at Nebraska. In 1930 the Ford Motor Company, Cork, Ireland, sent Fordson kerosene (Test 173) and the gasoline (Test 174) fueled tractors for official testing. More Ford tractors followed in 1937 and 1938. Since then, the Laboratory has tested tractors from Canada, England, France, Germany, Sweden, Poland, the Soviet Union, Japan, Czechoslovakia, Ireland, and Italy.

Many of the tractors manufactured by American companies now are built partially in other countries and use parts from foreign suppliers. Some of the same American companies also import tractors from other countries to be sold in the US.

Pneumatic Tire Development

In the 1920's it had become increasingly apparent that tractor drive wheel lugs were not compatible with hard surfaced roads. Solid rubber truck tires and later, high pressure pneumatic tires for trucks and cars did not function well under field conditions.

Mr. Hoyle Pounds, of Winter Garden, Florida, in 1928 mounted two pneumatic truck tires and rims on each side of the rear of a Fordson Tractor. No inner tubes were used; the heavy sidewalls of the truck tires supported the tractor and flexed as they revolved. The tires were bolted to the rims at four equally spaced points to prevent tire-to-rim slippage.

Mr. Pounds found that the maximum travel speed of the Fordson was too slow for optimum performance. He obtained a final drive worm and gear that practically doubled the rate of travel. This combination showed considerable promise in Florida's sandy field and grove conditions. The need for improved braking to control tractor and load was also evident.

Firestone, working with Allis-Chalmers, installed airplane tires on the Model U. Goodyear did the same on the Case Model C tractors.

These early efforts were so promising that both Firestone and Goodyear initiated design and development of agricultural tractor tires. Goodrich and US Rubber quickly followed and a test program was conducted near Phoenix, Arizona. An industry grant was made to the Agricultural Engineering Department, University of Nebraska where field tests were conducted and reported in Experiment Station Bulletin No. 291, published in 1934 by C. W. Smith and L. W. Hurlbut.

Early in 1934, the University of Nebraska Tractor Test Board offered to conduct drawbar tests on pneumatic rubber tires for no additional fee over the usual steel wheel tests. Applications were on file for Test No. 221—I H Farmall, No. 222—John Deere row crop and No. 223, Allis-Chalmers WC. Harvester and Deere declined; Allis-Chalmers enthusiastically accepted. The WC became the first tractor officially tested on rubber tires at the Nebraska Tractor Testing Laboratory.

The improvement in performance was remarkable when pneumatic rubber tractor tires were used during Test No. 223. During the two economy runs the horsepower showed over 25 percent increase at the same rpm and almost the same load. With steel wheels the tractor travelled in 2nd gear while the rubber tired tractor could operate with a similar load in 3rd gear. Fuel economy was improved over 25 percent. Rubber tires for tractors was surely a MILESTONE in the development of the agricultural tractor.

In this and succeeding tests, the need for additional weight on the drive wheels to reduce slippage was obvious. Larger tires both in overall diameter and in section, were also indicated.

Added weight was sometimes concrete poured around the spokes between rim and hub, particularly on field changeovers. Here the rims for lugs were cut away and tire rims welded in their place.

Tractor manufacturers offered cast iron wheel weights that could be bolted to the drive wheel and successfully

Allis-Chalmers was one of the early manufacturers of tractors to try pneumatic rubber tires on a farm tractor. Early experimental work was carried on with the cooperation of the Firestone Tire and Rubber Co. and Albert Schroeder, a Wisconsin farmer shown on his Allis-Chalmers Model "U".

attached to the previous weight until the recommended maximum load carrying capacity of the tire was reached. Something less than maximum weight usually produced optimum results. In the late 1930's, interest was shown in filling the drive wheel tires with liquid for added weight. Calcium chloride added to water (3-5 lbs./gal.) increased the weight of the solution and lowered its freezing point to an acceptable figure.

The ground-gripping projections on the tread of the first agricultural rear tractor tires were low, approximately ½ in. high and relatively closely spaced. Goodyear used their diamond design; Firestone a series of low chevrons.

In a few years tread design adopted by almost all tire manufacturers for farm field work had moved from the closely spaced patterns to relatively widely spaced bars, higher and heavier, still using approximately the same total amount of rubber in the tread. This has continued through the years, with tread bars changing as to height, width, angle and to some degree pattern. A continual striving for improved traction, wear, cleaning and body strength has been an industry objective. Special tread and body modifications have been developed for muddy rice field conditions, for forestry and for off-the-road operations. Tires for carrying heavy loads over soft ground were developed. The increase in power of agricultural tractors was accompanied by increases in tire and rim diameters, tire cross section and rim width. Improvement in tractive ability was accompanied with problems in tire side wall buckling and slippage of the tire bead on the rim. During this same period the tire section height in relation to the tire section width was decreased. Dual and even triple tires on the traction wheels of both two and four-wheel drive tractors are often used for improved traction and flotation. In an attempt to improve the performance of tractor tires, radial tractor tires had been introduced about 1970. Many European tractors had been equipped with radial tractor tires since late 1950.

How to read the specifications in this book

manufacturer: **Allis-Chalmers**

model number: **7030**

fuel: **Diesel**

Nebraska test number: **1119**

year or years built: **1973**

observed maximum belt or PTO and drawbar horsepower: **130.98 - 113.18 hp @ 2300 rpm,**

bore and stroke: **4¼" x 5";**

size, type of engine, engine manufacturer: **426 cu in. 6 cyl turbocharged Allis-Chalmers engine;**

gearing and speeds: **20 forward gears, 1.5 to 19 mph;**

fuel economy data during 10 hr run at 75% load: **11.90 hp hr/gal;**

weight without ballast: **12,430 lbs**

operator sound level at 75% load: **79.5 dB(A)**

FARM TRACTORS 1950-1975

1950

Horsepower Race Begins

The year 1950 ushered in a new decade for the agricultural tractor, marked by an increase in horsepower and the use of diesel and LPG fuel, improvements in engine lubrication, hydraulic lift systems, comfort and safety features for the operator. Suppliers were providing high altitude pistons and over-sized sleeves and pistons as replacements to increase power.

Farms were increasing in size and requiring more power. Deere and Company realized that a larger tractor than the Model D was needed. About the same time diesel fuel was becoming increasingly more popular for industrial use. The disadvantages of diesel fuel were being reduced as a result of new advances in diesel engines. It seemed more apparent that diesel power should be adopted to agricultural use.

Deere Model R

In 1949 the John Deere Waterloo Tractor Works introduced the Model R Diesel tractor. This supplemented the Model D tractor that had been introduced in 1923. The Model R was the largest and most powerful tractor that had been built in Waterloo.

The new Model R Diesel tractor engine was started by an auxiliary 2-cylinder gasoline engine that also assisted in warming up the engine for better starting in cold weather. A throw-out clutch was used to disengage the hydraulic lift system, when not needed, in order to con-

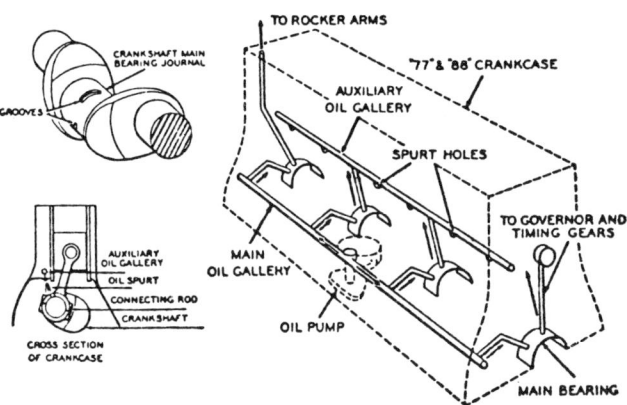

Engine lubrication system is a combination of the timed pressure and oil spurt type. Main bearings are pressure lubricated. Rod bearings are jet lubricated from the spurt holes in the auxiliary oil gallery.

This M&W Gear ad ran in Implement & Tractor magazine in 1950. An add-on attachment for faster travel speeds.

serve power. An optional, innovative feature of the Model R was a steel cab.

LPG Use Increases

The Kansas State College Engineering Experiment Station Bulletin No. 71, "Liquefied Petroleum Gas For Tractors" by George Larson, gave the following information about LPG as fuel for farm tractors: "Liquefied Petroleum Gas (LPG) has become an important fuel for

1950-1975 FARM TRACTORS

farm tractors. A number of manufacturers had provided factory equipped tractors using LPG fuel. However, not all these tractors had changed the basic design of the engines but merely changed the fuel system in such a way that it could use LPG fuel."

Kits to Convert to LPG

Farmers and dealers were able to obtain kits (from various sources) to convert their tractors to use LPG fuel in their present engines. Two types of conversions were in use: The vapor withdrawal system and the liquid withdrawal system. The liquid withdrawal system was usually referred to as the standard system and most often used by the factory equipped tractor. The vapor withdrawal system, which was cheaper to install, was often used in field conversions.

Either method, when used in field conversions, might utilize the original carburetor with a "spud in" gas discharge nozzle or adaptor for the gasoline carburetor. In other cases a special LPG carburetor might be used.

This shows how one farmer converted a Ford tractor to LPG—many farmers made these conversions.

In this conversion the regular gasoline tank was replaced with a LP tank.
Photos courtesy George Larson, Kansas State University.

Minneapolis-Moline LPG Tractors

1929 was the year that the Minneapolis-Moline Power Implement Company first came into existence as a result of the merger of Moline Plow Company, founded in 1870; the Minneapolis Threshing Machine Company, organized in 1887; and Minneapolis Steel and Machinery Company, 1902. Its name was shortened to Minneapolis-Moline Company in 1949, and was changed several times after that until it became a wholly-owned subsidiary of the White Motor Corporation in 1963.

Minneapolis-Moline has been recognized as pioneers in the design of tractors and power units using Liquefied Petroleum Gas (LPG) for agricultural work. In 1941, Minneapolis-Moline started to release tractors and power units equipped to utilize LPG. In 1949 the Model U LPG (Test 441) was the first factory designed tractor to be tested at the Nebraska Tractor Testing Laboratory using LPG as a fuel.

Minneapolis-Moline followed a design which limited the engine speed on all their tractors and power units below the usual speed increases used for many years by their competitors.

The early Minneapolis-Moline tractors used a swing seat providing room for the operator to easily stand on the tractor platform while driving the tractor. Later, the seat was covered with an air cushion and a padded back, for more comfort. A hand clutch lever was used until 1956.

Model Z

The Model Z tractor was introduced about 1950. This tractor had a very unusual engine design with the valves operating horizontally by means of vertically mounted rocker arms driven directly from the cam shaft (without push rods). One of the features advertised was the ease with which the operator could adjust the valves from a comfortable sitting position. In this same position, it was also possible to work on the crankshaft bearings. Another unusual feature of this tractor engine was the mounting of the crankshaft on ball bearings at each end and an oil pressure ring at the center of the crankshaft. Oil was supplied under pressure to the connecting rods by means of the drilled crankshaft.

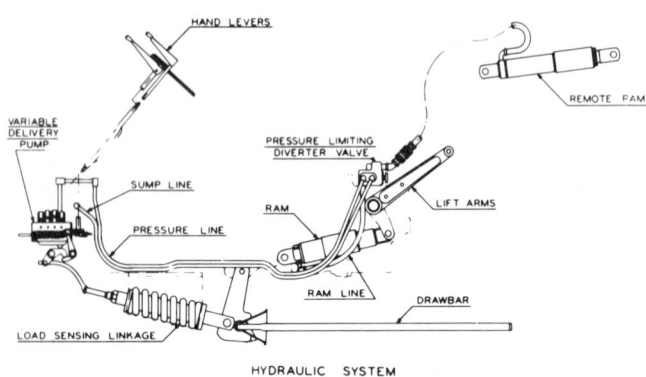

The Traction Booster system utilizing the hydraulics to control the draft of the implements. This was a new feature of the Allis-Chalmers WD introduced in 1948.

Traction Booster Used

The new Allis-Chalmers CA tractor was similar to its Model B but developed more power, because of a higher compression ratio and increased engine rpm. This tractor also had a "Traction Booster" system which used the hydraulics to control implements pulled by the tractor. A 2-clutch power control provided continuous PTO operation. The operator used a foot clutch for drawbar work and a hand clutch to permit the power take-off to operate when the tractor was not moving.

Developments at Oliver

Oliver engines of this period featured an engine lubrication system which included a constant pressure circuit and a spurt circuit. Main bearings were pressure-lubricated and rod bearings jet-lubricated from spurt holes in the auxiliary oil gallery.

The Oliver 88 was the first of Oliver's New Fleetline tractor Series which included the 88, 77, and 66 models superseding the 80, 70 and 60 Series. The New 88 Series was introduced in connection with the 100th anniversary (1948) of Oliver. Standardization of many engine and tractor parts was emphasized.

The new series were all available with gasoline and kerosene-distillate (KD) engines until 1949. Following 1949 all the New Series tractors were available with Oliver designed diesel engines replacing the KD Models. Beginning in 1937 the first Oliver Row Crop Diesel Model 80 was available.

Massey-Harris Tractors Increase in Size

The first Massey-Harris tractors were tested at Nebraska in 1930. Between 1930 and 1950 23 models were tested.

The Massey-Harris 44 and 55 Models were quite popular following the war period. The Massey-Harris 55 Diesel was one of the largest farm tractors produced by this company and was one of the largest farm wheel tractors on the market at that time.

The Model 55 was available to use gasoline, diesel, LPG and low grade ("K") fuel.

The Models 44 and 55 were equipped with hydraulic capabilities to operate remote hydraulic cylinders on trailed implements.

Cockshutt and Oliver

Prior to WW II the Cockshutt Co., a Canadian company founded in 1877, had been marketing tractors in Canada built by the Oliver Corporation plant at Charles City, Iowa. These tractors were painted a different color for the Canadian market. Following the war, Cockshutt developed a line of tractors in Brantford, Ontario. These tractors were distributed in the US by Gambles Stores and the Co-op. Seven Cockshutt tractors were tested at Nebraska between 1947 and 1958. The Cockshutt 30 (Test 382) was the first tractor with *independent* PTO to be tested at Nebraska.

Others Bid For Market

In addition to the widely-known manufacturers of agricultural tractors, other companies were making their bids for a market with innovations of their own. The Lodge and Shipley Company, Cincinnati, was producing the Choremaster, a tractor with 0.77 horsepower on the drawbar and an engine which developed a maximum of 1.47 horsepower (Test 449).

The Willys Farm Jeep, a 4-wheel truck-tractor vehicle equipped with a 3-point hitch, hydraulic lift, and a PTO at the rear was offered for sale as a farm tractor.

The Intercontinental Manufacturing Company, Grand Prairie, Texas, began building tractors for the export market. Several thousand Intercontinental tractors were built and sold in Cuba, Mexico, India, and Argentina. The company produced tractors for only five years.

Allis-Chalmers B Gasoline   439
1946-1957
22.25-19.51 hp @ 1500 rpm, 3 3/8" x 3 1/2";
125.3 cu in. 4 cyl Allis-Chalmers engine;
3 forward gears, 2.4 to 7.5 mph;
9.86 hp hr/gal; 2251 lbs.

Allis-Chalmers CA Gasoline   453
1950-1957
25.96-22.97 hp @ 1650 rpm, 3 3/8" x 3 1/2";
125 cu in. 4 cyl Allis-Chalmers engine;
4 forward gears, 2 to 11.3 mph;
10.29 hp hr/gal; 2763 lbs.

1950-1975 FARM TRACTORS

Allis-Chalmers WD Gasoline 440
1948-1953

34.63-30.23 hp @ 1400 rpm, 4'' x 4'';
201 cu in. 4 cyl Allis-Chalmers engine;
4 forward gears, 2.2 to 9.6 mph;
9.97 hp hr/gal; 3721 lbs.

Test 399 in 1948 Model WD Distillate with 201 cu in. displacement and same rpm developed 26.14 hp and 9.44 hp hr/gal.

Choremaster B Gasoline 449
1950

1.47-0.77 hp @ 3600 rpm, 2'' x 1 7/8'';
5.89 cu in. 1 cyl air-cooled Clinton engne;
1 forward gear, 1 3/4 to 2 2/3 mph;
3.30 hp hr/gal; 123 lbs.

Smallest tractor ever tested at Nebraska.

Cockshutt 30 Gasoline 442
1950

43.30-37.85 hp @ 1650 rpm, 3 7/16'' x 4 1/8'';
230 cu in. 6 cyl Buda engine;
6 forward gears, 1.6 to 12.0 mph;
9.72 hp hr/gal; 5305 lbs.

First tractor tested at Nebraska equipped with independent PTO (live PTO with independent clutch).

John Deere R Diesel 406
1949-1954

48.58-43.15 hp @ 1000 rpm, 5 3/4'' x 8'';
415.5 cu in. 2 cyl John Deere engine;
5 forward gears, 2.1 to 11.5 mph;
15.19 hp hr/gal; 7603 lbs.

Unusually good fuel economy—first Nebraska Test to record 0.400 lbs/hp hr on belt test and 15.19 hp hr/gal during drawbar tests. Used a 2-cylinder horizontal opposed gasoline engine for starting. This engine had a 2 6/10'' x 2 5/16'' bore and stroke running at 4000 rpm.

FARM TRACTORS 1950-1975

Dodge Power Wagon T-137 Gasoline   454
1950
42.40-40.02 hp @ 1700 rpm, 3 1/4" x 4 3/8";
230.2 cu in. 6 cyl Chrysler "L" Head engine;
8 forward gears, 2 to 54 mph;
7.05 hp hr/gal; 5809 lbs.

Ford 8-N Gasoline   443
1947-1952
26.19 hp @ 2000 rpm/21.95 hp @ 1750 rpm, 3 3/16" x 3 3/4";
119.7 cu in. 4 cyl Ford "L" Head engine;
4 forward gears, 3.2 to 11.9 mph;
10.13 hp hr/gal; 2717 lbs.

Model 8 NAN Distillate (Test 444) with same displacement and rpm developed 21.87 hp and 8.80 hp hr/gal.

Massey-Harris 55 Diesel   452
1949-1955
59.04-52.49 hp @ 1350 rpm, 4 1/2" x 6";
382 cu in. 4 cyl Continental engine;
4 forward gears, 3 to 12 mph;
13.95 hp hr/gal; 7793 lbs.

The second diesel-powered Massey-Harris tractor. In 1948, the Model 44 Diesel (Test 426) was introduced with 260 cu in. engine. In 1951, the Model 55 Gasoline (Test 455) with same displacement and rpm developed 66.91 hp and 10.38 hp hr/gal.

Minneapolis-Moline G Gasoline   437
1947-1951
58.03-49.53 hp @ 1100 rpm, 4 5/8" x 6";
403.2 cu in. 4 cyl Minneapolis-Moline engine;
5 forward gears, 2.5 to 13.8 mph;
9.72 hp hr/gal; 7230 lbs.

1950-1975 FARM TRACTORS

Minneapolis-Moline U Standard LPG 411
1947-1959
46.87-41.05 hp @ 1275 rpm; 4 1/4" x 5";
283 cu in. 4 cyl Minneapolis-Moline engine;
5 forward gears, 2.7 to 14.8 mph;
7.56 hp hr/gal; 5979 lbs.

In 1949 first LPG tractor was tested at Nebraska.

Oliver 88-HC Gasoline 388
1947-1952
41.99-36.97 hp @ 1600 rpm, 3 1/2" x 4";
230.9 cu in. 6 cyl Oliver engine;
6 forward gears, 2.6 to 12.3 mph;
9.47 hp hr/gal; 5285 lbs.

Oliver Row Crop 77-HC Gasoline 404
1948-1954
33.98-28.68 hp @ 1600 rpm, 3 5/16" x 3 3/4";
193.9 cu in. 6 cyl Oliver engine;
6 forward gears, 2.6 to 12.25 mph;
8.82 hp hr/gal; 4828 lbs.

Oliver Standard 77 HC Gasoline (Test 405) essentially the same.

Oliver Standard 77-HC was retested (425) in 1949 with same displacement and rpm developing 37.17 hp and 9.90 hp hr/gal. Model Row Crop 77 LPG (Test 470) with same displacement and rpm developed 36.33 hp and 7.32 hp hr/gal.

Oliver 88 Diesel 450
1950-1954
43.53-38.30 hp @ 1600 rpm, 3 1/2" x 4";
230.9 cu in. 6 cyl Oliver engine;
6 forward gears, 2.5 to 12 mph;
12.91 hp hr/gal; 5680 lbs.

First Oliver agricultural tractor tested at Nebraska using a diesel engine.

FARM TRACTORS 1950-1975

1951

Seats Change For Better

Before the 1950's, tractor manufacturers paid little attention to the comfort of tractor operators. Many tractors still are equipped with hard, pressed steel pan seats, although some seats were covered with padding. Now things were changing for the better. Shock absorbers, coil springs and other devices were making their appearance to take out some of the discomfort.

Minneapolis-Moline tractors used a swing seat, providing room for the operator to stand on the tractor platform while driving. Later the seat was covered with an air cushion and a padded back was added. Allis-Chalmers used a coil spring combined with a shock absorber to support the seat on its WD Model. International Harvester was working on improvements with the upholstered seat on its Farmall C. The seat had a conical variable-rate spring suspension and hydraulic shock absorber to ease up-and-down motions when the tractor was traveling over rough terrain.

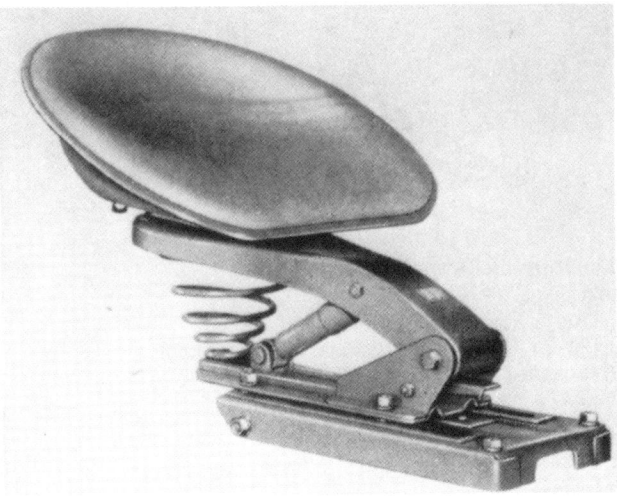

International Harvester provided an upholstered seat with adjustable double-action shock absorber and a conical spring on its 1951 models.

Many manufacturers were increasing horsepower ratings in 1951. International Harvester changed the model designation of its Farmall C to Farmall Super C (Test 458), increasing belt horsepower from 21.1 to 23.7. But the horsepower surge was just beginning. Most tractors tested at Nebraska between 1947 and 1951 were between 20 and 30 horsepower.

The Ferguson Hitch

The Ferguson TE-20 tractor was first built in 1946 by the Standard Motor Company in England. The US version, known as the TO-20 was produced at the Ferguson Park Plant in Detroit, Michigan. The first TO-20 was produced October 11, 1948. The Ferguson TO-30, an improved version of the TO-20, appeared in 1951.

While there was much skepticism about the 3-point Ferguson system at the beginning in 1939, it later (1959) became the Category I standard that was adopted by the American Society of Agricultural Engineers and the Society of Automotive Engineers.

The Farmall Super C was equipped with this disc brake. It used tractor motion to apply pressure to the larger braking area. The brake was easily adjusted and sealed in a dust-proof and moisture-proof housing.

Disc Brakes Appear

About this time the disc brakes began to appear as standard equipment on some tractor models. The brakes on the Farmall Super C utilized the tractor's motion to apply pressure to their larger braking area. They were easily adjusted and sealed in a dust-proof and moisture-proof housing.

Brakes on the Ferguson TO-30 were operated from both sides of the tractor. This made it necessary for the operator to use his left foot for both the brake and the clutch on the left side of the tractor. However, there was also a master brake pedal on the right side to operate both brakes.

Oliver started using disc brakes on its Model 88. In general, disc brakes were an improvement over contracting band brakes. They provided longer life and less fade.

Some band and shoe brakes did not work well when the tractor traveled in reverse.

Minneapolis-Moline

Minneapolis-Moline and B. F. Avery & Sons merged. The Avery tractor line was discontinued within a very few years.

The Model Z tractor had been restyled somewhat and was now known as the Model ZA with same performance as the original Z.

The Model R was announced in 1948 equipped with disc brakes. This tractor had many of the same engine features as used on the Model Z. The disc brakes provided longer life than the original band or brake shoes used earlier.

The Model L Uni-tractor was introduced in 1952. This tractor was designed to accept several pieces of equipment, such as the Uni-combine, Uni-husker, Uni-picker/sheller, Uni-baler and Uni-forager. The Model L originally had a V-4, 206 cu in. displacement engine which was later (1960) replaced with a 206 cu in. displacement in-line engine. The Uni-tractor and equipment line were sold to New Idea in 1962.

McCormick Farmall Super C Gasoline   458
1951-1954
23.67-20.72 hp @ 1650 rpm, 3 1/8" x 4 ";
122.7 cu in. 4 cyl International Harvester engine;
4 forward gears, 2.4 to 10.25 mph;
8.88 hp hr/gal; 3209 lbs.

Replaced McCormick Farmall C.

Ferguson TO-30 Gasoline   466
1951-1954
29.32 hp @ 2000 rpm/24.37 hp @ 1750 rpm, 3 1/4" x 3 7/8";
129 cu in. 4 cyl Continental engine;
4 forward gears, 2.9 to 11.5 mph;
9.98 hp hr/gal; 2843 lbs.

Drawbar tests made at 1750 rpm. Replaced Model TE-20.

McCormick Farmall WD-6 Diesel   459
1951-1952
37.64-33.25 hp @ 1450 rpm, 3 7/8" x 5 1/4";
247.7 cu in. 4 cyl International Harvester engine;
5 forward gears, 2.4 to 15.75 mph;
12.23 hp hr/gal; 5789 lbs.

Minneapolis-Moline BF Avery Gasoline 469
1951
27.12-24.12 hp @ 1800 rpm, 3 1/4" x 4";
133 cu in. 4 cyl Hercules engine;
4 forward gears, 2.4 to 13.1 mph;
9.10 hp hr/gal; 2894 lbs.

Oliver 66 Diesel 467
1951-1954
25.03-22.05 hp @ 1600 rpm, 3 5/16" x 3 3/4";
129.3 cu in. 4 cyl Oliver engine;
6 forward gears, 2 to 12 mph;
12.09 hp hr/gal; 3795 lbs.

Minneapolis-Moline R Gasoline 468
1951
27.09-23.90 hp @ 1500 rpm, 3 5/8" x 4";
165 cu in. 4 cyl Minneapolis-Moline engine;
4 forward gears, 2.6 to 13.2 mph;
7.95 hp hr/gal; 3414 lbs.

One of the first Minneapolis-Moline tractors equipped with disc brakes.

1950-1975 FARM TRACTORS

1952

Horsepower Goes Up

By the second year of the new decade, the race for increased horsepower was heating up, improvements in performance were well underway, and power steering was making its debut.

The Case LA tractor was introduced during the early 1940's replacing two earlier models. The Model L Kerosene (Test 155) had been in service since 1929 until it was replaced by the Model L Distillate (Test 309) with a slight increase in power. The Model LA was not officially tested until 1952 after some improvements had been made increasing the power still more. The Model LA was available not only as a distillate tractor but also as a gasoline and LPG model.

The enclosed roller chain final drive, which ran in oil, was carried over from the earlier Model L. A hand lever was used to operate a wet clutch, while the magneto was to be used for the ignition system for a few more years.

Operator comfort was being considered by the early use of a weatherproof covered sponge rubber cushion placed on the pressed steel seat. It was possible for the operator to swing the seat from side to side permitting him to stand more conveniently on the platform and also to get on and off the tractor easier.

The Case Eagle Hitch, introduced in 1949 on the Model VAC was now available on D- and S-series tractors, and remote hydraulic control circuits were available on most Case models. The Eagle Hitch was a depth control 3-point hitch system with laterally rigid draft arms. "Claws" on the draft arm ends could be unlatched to aid the operator in hooking up to an implement without leaving his seat. They would lock automatically as the tractor moved forward and the implement was raised after hookup. Versions of the Eagle Hitch were used on many later models, including the 300, 400, 500 and 600 Rock Island-built models, the 700, 800, 730, 830, and others, until it was discontinued in 1964.

International Harvester changed its model designations on the Farmall series to Farmall Super M, Farmall Super MD, McCormick Super W-6 and McCormick Super WD-6. The belt horsepower was increased from 35 to 45. The same year International introduced its Super M LPG.

First Minneapolis-Moline Diesel

The first Minneapolis-Moline diesel tractor was announced about 1952. This was the Model U tractor. This tractor had the same 283 cu in. displacement engine and rpm as the original gasoline Model U. A very limited number of these tractors were built.

World Market For Fordson

Ford was producing its Fordson Major at the Dagenham plant in the United Kingdom. By 1964 the Major was to become Ford's most widely exported tractor, establishing the company in many world markets. The 3-point hitch used on the Fordson Major was later adopted as an ASAE standard for the Category 11 hitch in 1959.

John Deere began using a numbered model designation on its Model 50 and 60 tractors. These new models replaced the Model B and Model A respectively. Improvements over the older models included a separate carburetor for each cylinder for better fuel distribution. These were the first John Deere tractors to use gasoline, All Fuel (distillate) or LPG.

This also was the year in which John Deere introduced its "live" power take-off. It enabled continuous power take-off operation by means of a separate control lever whenever the engine was running. An independent hand clutch provided direct drive from the engine to the PTO. Wheel spacing was achieved by means of rack and pinion

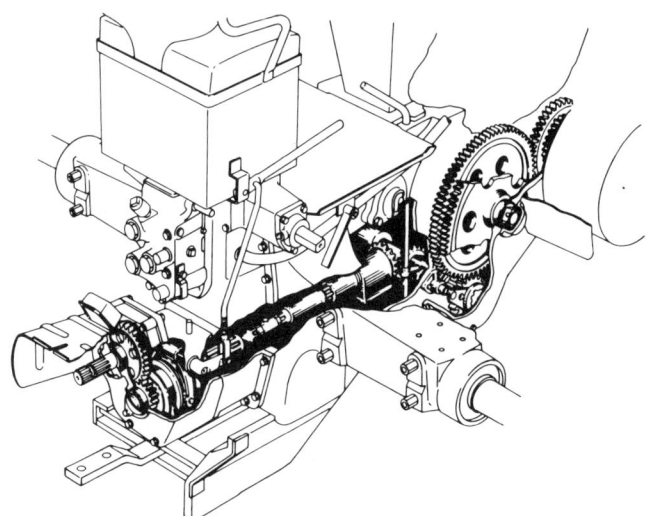

This is the PTO drive used on Models 50, 60, and 70 John Deere tractors introduced in 1952 and 1953. An independent hand clutch provided direct drive from the engine to the PTO.

1950-1975 FARM TRACTORS

In 1952 the Behlen Manufacturing Company made available a power steering system which farmers could install themselves. It included a hydraulic pump powered by the tractor engine, a hydraulic gear motor coupled to the steering wheel shaft and the necessary hoses and fittings. (Courtesy Walt Behlen Universe)

adjustment. A cushioned seat with back rest improved operator comfort.

Farmers Welcome Power Steering

After fighting their steering wheels for years in an effort to keep their tractors on a true course across bumpy fields, farmers welcomed power steering.

The Behlen Manufacturing Company of Columbus, Nebraska made available a power steering system which farmers could install themselves. It included a hydraulic pump powered by the tractor engine, a hydraulic motor coupled to the steering wheel shaft, and the necessary hoses and fittings. The unit sold for $180 FOB Columbus, Nebraska.

As the operator of the agricultural tractor began to enjoy the improved comforts of riding and operating his machine, engineers in tractor manufacturing plants and agricultural engineers and extension specialists at universities throughout the nation turned increasing attention to the operator's safety. The number of accidents involving farm tractor operators was increasing. The machines with more power, speed and maneuverability than before, had a way of tipping over when misused.

Safety Demonstrations Staged

Manufacturer representatives and university engineers began staging tractor safety demonstrations in many rural communities. One such demonstration took place at the University of Nebraska's Tractor Power and Safety Day. Viewers watched a tractor, protected by a special

The safety frame on the John Deere tractor was intended to protect the tractor, but not the driver "Jughead". This was an early "safety frame" for tractor tip over.

safety frame and operated by remote control, perform various tips, rolls, and other mishaps due to faulty operation or adjustments. "Jughead", the dummy sitting in the tractor seat, was always a casualty, but he continued to play his role nearly 450 times in the next 25 years.

During this same period, kits were available to provide other measures intended to protect tractor operators. Some suppliers placed switches under the tractor seat which would shut off the engine ignition when the operator left the seat. Mercury switches also were used to shut off the engine when the tractor approached a certain degree of slope. However this was usually too late to prevent a tip over.

J.I. Case LA Gasoline   480
1952-1953
58.53-51.68 hp @ 1150 rpm, 4 5/8" x 6";
403.2 cu in. 4 cyl Case engine;
4 forward gears, 2.5 to 10 mph;
10.24 hp hr/gal; 7621 lbs.

Model LA Distillate (Test 481) with same displacement and rpm developed 48.86 hp and 9.34 hp hr/gal.

Model LA LPG (Test 482) with same displacement and rpm developed 59.60 hp and 8.14 hp hr/gal.

Cockshutt 20 Gasoline   474
1952
28.94-25.47 hp @ 1800 rpm, 3 3/16" x 4 3/8";
140 cu in. 4 cyl Continental engine;
4 forward gears, 2.5 to 13.25 mph;
9.17 hp hr/gal; 2813 lbs.

1950-1975 FARM TRACTORS

Cockshutt 50 Gasoline   488
1952
55.56-51.59 hp @ 1650 rpm, 3 3/4" x 4 1/8";
273 cu in. 6 cyl Buda engine;
6 forward gears, 1.5 to 9.85 mph;
10.69 hp hr/gal; 6041 lbs.

Model 50 Diesel (Test 487) with same displacement and rpm developed 51.05 hp and 12.80 hp hr/gal.

John Deere 60 Gasoline   472
1952-1956
40.24-35.18 hp @ 975 rpm, 5 1/2" x 6 3/4";
321 cu in. 2 cyl John Deere engine;
6 forward gears, 1.5 to 11 mph;
10.22 hp hr/gal; 5911 lbs.

In 1953, Model 60 All Fuel or Tractor Fuel (Test 490) with same displacement and rpm developed 32.48 hp and 9.41 hp hr/gal.

In 1953, Model 60 Gasoline (Test 513) with same displacement and rpm developed 41.18 hp and 7.96 hp hr/gal.

John Deere 50 Gasoline   486
1952-1956
30.35-27.07 hp @ 1250 rpm, 4 11/16" x 5 1/2";
190.4 cu in. 2 cyl John Deere engine;
6 forward gears, 1.5 to 10 mph;
10.51 hp hr/gal; 4855 lbs.

In 1953, Model 50 All Fuel (Test 507) with same displacement and rpm developed 24.83 hp and 10.07 hp hr/gal.

In 1955, Model 50 LPG (Test 540) with same displacement and rpm developed 31.20 hp and 7.83 hp hr/gal.

McCormick Super WD-6 Diesel   478
1954-1956
46.84-41.77 hp @ 1450 rpm; 4" x 5 1/4";
264 cu in. 4 cyl International Harvester engine;
5 forward gears, 2.6 to 16.1 mph;
12.77 hp hr/gal; 6011 lbs.

Model Super W-6 Gasoline (Test 476) same displacement and rpm developed 46.25 hp and 9.87 hp hr/gal.

Model Super W-6 LPG (Test 485) same displacement and rpm developed 47.53 hp and 7.62 hp hr/gal.

McCormick Farmall Super MD Diesel 477
1952-1954
46.73-42.19 hp @ 1450 rpm, 4" x 5 1/4";
264 cu in. 4 cyl International Harvester engine;
5 forward gears, 2.6 to 16.75 mph;
13.13 hp hr/gal; 6034 lbs.

Replaced Model MD Diesel.

Model Super M Gasoline (Test 475) same displacement and rpm developed 46.26 hp and 10.30 hp hr/gal.

Model Super M LPG (Test 484) same displacement and rpm developed 47.07 hp and 7.83 hp hr/gal.

New Fordson Major Diesel 500
1952-1961
38.49-34.17 hp @ 1600 rpm, 3.938" x 4.524";
220 cu in. 4 cyl Ford (England) engine;
6 forward gears, 2.07 to 13.16 mph;
13.93 hp hr/gal; 5308 lbs.

The Fordson Major Gasoline (Test 501) with 199 cu in. displacement and same rpm developed 33.56 hp and 8.19 hp hr/gal.

1950-1975 FARM TRACTORS

1953

Turbochargers Answer Demand For More Power

More horsepower—that was the battle cry of tractor users as the industry moved into 1953. And the urge for more power did not go unnoticed. Suppliers began to provide accessories, such as turbochargers, to increase power. By installing a turbocharger on a conventional diesel engine and increasing the fuel flow, it was possible to increase maximum horsepower considerably.

Many dealers would increase the fuel flow of diesel engines usually at the request of the user or buyer. Some operators increased engine rpm above the rated speed to gain power. Others installed high altitude pistons raising the compression ratio. Over-sizing the engine with sleeves and pistons provided for that purpose seemed to help satisfy the urge for more power. Still others tried larger engines as replacements for the original engine. It wasn't long until some tried injecting propane into the air intake of their diesel tractors in a desire for more power.

All these practices disturbed manufacturers because it often over-taxed the design of the engines and drive train.

New Models More Powerful

The J.I.Case Company introduced the new Model 500 diesel tractor. This was the first agricultural diesel and the first 6-cylinder engine tractor produced by the Racine Tractor Plant. This new model resembled the Case LA, which it replaced, with at least a 10 percent increase in power. Other new innovations included the first Case power steering for operator convenience. In addition the Model 500 provided optional independent PTO with a separate clutch control. The cushioned operator's seat was supported by rubber in torsion and could be tilted on an angle to provide rainwater drainage or greater standing room.

The production of the last Model SC tractor in 1954 marked the decline of distillate engines by Case. Diesel engines were gaining in popularity and diesel fuel was gradually replacing other fuels for future farm tractors.

The announcement of the Model 500 Diesel in 1953 was also the start of new model designation system at Case. The first digit in the identification number listed the series number. The second digit indicated the type of fuel used. For example, on some models (0) meant diesel fuel; (1) meant spark ignition for gasoline or LPG; and (2) meant Industrial. The third digit indicated the type of tread. For example, (0) meant standard fixed tread; (1) indicated a General Purpose tractor; (2) indicated special tractors for orchard, western or rice use; (3) represented high clearance. Variations of this method of identification continued until 1969.

In 1953 Case produced the last of its famous threshing machines which had been in use for many years. The "combine" had caused the demise of a long line of successful Case threshers.

Massey-Harris and Ferguson Merge

On August 12, 1953 Massey-Harris and Ferguson merged. Negotiations had started in 1947. In March of 1954 the new entity adopted the new name of Massey-Harris-Ferguson.

In tractor production, the Massey-Harris 33 gasoline and diesel models replaced the earlier Model 30, Nebraska tested in 1949. The Massey-Harris 44 Special replaced the earlier Model 44 with additional power. The tractor was also available as a diesel.

This was the year Ford chose to introduce its NAA tractor. It was called the Golden Jubilee, in recognition of the 50th anniversary of the Ford Motor Company. Ford also opened offices and a new farm machinery center near Birmingham, Michigan, this year.

Allis-Chalmers came out with its WD 45, available for use with four different fuels—gasoline, distillate, diesel, and LPG. The new model had nine horsepower more than the Model WD it replaced. In order to have the same power as the gasoline and LPG tractors, the diesel power unit used a 6-cylinder engine. This was the first 6-cylinder engine used in any Allis-Chalmers agricultural tractor. Another innovation introduced on the WD 45 was a snap-coupler hitch which permitted the operator to couple and uncouple implements from the tractor seat.

International Harvester changed the model designation of its Farmall H and McCormick W-4 to Farmall Super H and McCormick Super W-4. It also added "Super" to the McCormick WD-9, with a corresponding increase in belt horsepower from 51.3 to 65.2.

For its McCormick Farmall Super C, International Harvester introduced a 2-point "fast hitch." The new-type hitch was provided in order to better compete with the 3-point hitch which was becoming popular among tractor users.

1950-1975 FARM TRACTORS

The new Allis-Chalmers WD-45 featured the Snap-Coupler hitch which permitted the operator to couple and uncouple implements from the tractor seat. This photo shows the Snap-Coupler hitch and the lift arms as used until 1967. The 3-point hitch with draft control was used on Allis-Chalmers tractors after 1967.

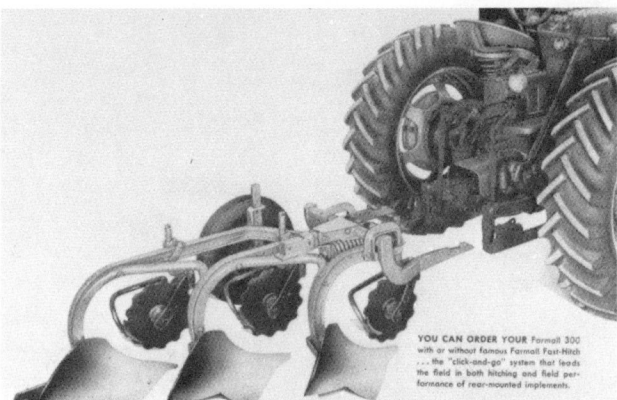

The Farmall 2-point "fast hitch" was designed to compete with 3-point hitch offered by other tractor manufacturers.

New Big Row Crop

John Deere introduced its largest row crop tractor to date—the John Deere 70. It was available with engines which could use gasoline, LPG, distillate, or diesel fuel. Other new features included optional power steering using a built-in hydraulic system and a mechanical weight-transfer system, the latter designed to transfer drawbar resistance into downward pressure on the rear wheels.

John Deere's Model 40 Standard and its 40 Tricycle replaced the M and MT tractors. These new models developed about 15 percent more power than the older tractors. They also carried an improved 3-point hitch with live hydraulic control.

On the foreign scene, David Brown Tractors, Ltd., Huddersfield, Yorkshire, England, exported its first tractor to the US and submitted it for test. The David Brown Model 25 was a gasoline tractor (Test 530).

While tractor manufacturers were busy turning out their new models, a historic event was taking place in Lincoln, Nebraska at the Tractor Testing Laboratory. As a result of a meeting between representatives of the tractor industry and members of the Tractor Test Board, the Nebraska Tractor Test Procedure was combined with the SAE tractor test code. This merger brought about much better understanding and cooperation between Nebraska Tractor Test Board and industry. It was the first step towards the adoption of a worldwide test code for all tractors.

Revival of Four-Wheel Drive

After 16 years of development work the Wagner Brothers of Portland, Oregon announced their first 4-wheel drive tractor in 1953. By 1956 Wagner Tractor, Inc. had really revived the interest in 4-wheel drive tractors for agriculture along the west coast and the great plains areas. There were three sizes available: The Wagner TR6 with a 57 hp engine, the Wagner TR9 with a 73 hp engine and the Wagner TR14 with a 96 hp engine. In 1957, Wagner tested a tractor for the first time. It was the TR9 diesel (Test 631). In fact, this was the first 4-wheel drive tractor to be tested since WW II except for the Willys Jeep, the Land Rover and the Dodge Power Wagon. By 1964, the Wagner TR14 (Test 700) and the Wagner FWD WA4 (Test 864) were subsequently tested.

Hal Higgins, a free lance writer in California, had been following the development of the Wagner tractors and wrote in *Farm Implement News*, July 10, 1956: "A 4-wheel drive tractor—The Wagner, is making good, chasing crawler tractors out of some prior-preempted fields. Others without the key hinge failed to make good. Remember the Fitch, the Nelson, and the Massey-Harris 4-wheel drive and others whose names fade from memory."

Prior to 1950 very few 4-wheel drive tractors were available except the Willys Jeep (Test 432), the Dodge Power Wagon (Test 454) and the English Land Rover (Test 749). Four-wheel drive tractors were not new and Gray had listed several in the "AGRICULTURAL TRACTOR: 1855-1950." A number of these 4-wheel drive tractors, all using steel wheels, are listed as follows:

YEAR	NAME	NEBR TEST NO.	REMARKS
1912	Olmstead	None	
1912	Nelson	None	
1920	Morton Four-Wheel Drive	None	
1922	Wilson	None	
1922	Rogers	84	First 4-wheel drive tractor tested at Nebr. Articulated with power steering
1926	Wizard	None	
1929	Fitch Four Drive	165 & 172	
1930	Massey-Harris Gen'l Purpose	177	
1931	Massey-Harris Gen'l Purpose	191	

Since 1936 all 4-wheel drive tractors have been equipped with pneumatic rubber tires which have scored a victory of sorts for Wagner.

In February 1961 the FWD Corporation of Clintonville, Wisconsin, acquired Wagner Tractor Inc. of Portland, Oregon. The new firm was named FWD Wagner Inc. and operated as a wholly owned subsidiary of FWD Corporation with management, manufacturing and sales out of the Portland office. In 1967-1968 Deere and Co. contracted to sell the Series 14 and 18 model tractors. Soon after this FWD Wagner discontinued selling for the agricultural market.

David Brown 25 Gasoline   530
1953-1958
34.89-26.88 hp @ 2000 rpm, 3 1/2" x 4";
154 cu in. 4 cyl David Brown engine;
6 forward gears, 1.5 to 13.9 mph;
9.59 hp hr/gal; 3731 lbs.

The first David Brown tractor tested at Nebraska.

Allis-Chalmers WD-45 Gasoline   499
1953-1957
43.21-37.84 hp @ 1400 rpm, 4" x 4 1/2";
226 cu in. 4 cyl Allis-Chalmers engine;
4 forward gears, 2.5 to 11.3 mph;
10.64 hp hr/gal; 3955 lbs.

Model WD-45 distillate (Test 511) with same displacement and rpm developed 33.01 hp and 9.23 hp hr/gal.

This was the last Allis-Chalmers tractor tested using distillate fuel.

Model WD-45 LPG (Test 512) with same displacement and rpm developed 44.13 hp and 8.34 hp hr/gal.

Model WD-45 Diesel (Test 563) with a 6-cylinder 230 cu in. engine at 1625 rpm developed 43.29 hp and 12.52 hp hr/gal. First 6-cylinder diesel engine in an Allis-Chalmers agricultral tractor.

J.I. Case SC Gasoline   496
1953-1954
31.71-27.68 hp @ 1600 rpm, 3 5/8" x 4";
165.1 cu in. 4 cyl Case engine;
4 forward gears, 2.5 to 10.3 mph;
10.28 hp hr/gal; 5007 lbs.

The Model SC Distillate (Test 497) with same displacement and rpm developed 24.97 hp and 9.35 hp hr/gal.

First Case tractors tested at Nebraska with independent PTO.

These tractors had steel pan seats with weatherproof cushions.

1950-1975   FARM TRACTORS

J.I. Case 500 Diesel   508
1953-1956
63.81-56.32 hp @ 1350 rpm, 4" x 5";
377 cu in. 6 cyl Case engine;
4 forward gears, 2.7 to 10.1 mph;
14.34 hp hr/gal; 8128 lbs.

Seat made of pressed steel (with sponge rubber cushion) swung from side to side and tilted upward.

First Case diesel tested at Nebraska and the first Case tested with power steering.

John Deere 70 Gasoline   493
1953-1956
48.29-42.24 hp @ 975 rpm, 5 7/8" x 7";
379.5 cu in. 2 cyl John Deere engine;
6 forward gears, 2.5 to 12.5 mph;
10.33 hp hr/gal; 6617 lbs.

Model 70 All Fuel (Test 506) with 412.5 displacement and same rpm developed 43.15 hp and 9.94 hp hr/gal.

Model 70 LPG (Test 514) with same displacement and rpm as the gasoline model developed 50.86 hp and 8.34 hp hr/gal.

In 1954, Model 70 Diesel (Test 528) with 376 cu in. displacement developed 50.40 hp at 1125 rpm and 15.78 hp hr/gal.

John Deere 40 Gasoline   503
1953-1955
24.25-21.71 hp @ 1850 rpm, 4" x 4";
101 cu in. 2 cyl John Deere engine;
4 forward gears, 1.6 to 12 mph;
9.38 hp hr/gal; 3219 lbs.

Model 40-S (Test 504) is essentially the same except it used a standard tread.

Model 40-S All Fuel (Test 546) with same displacement and rpm developed 20.04 hp and 8.69 hp hr/gal.

All Dubuque built tractors used foot clutches.

Ford NAA "Jubilee" Gasoline   494
1952-1954
31.14 hp @ 2000 rpm/25.30 hp @ 1750 rpm, 3.4" x 3.6";
134 cu in. 4 cyl Ford engine;
4 forward gears, 3.1 to 11.6 mph;
10.08 hp hr/gal; 2841 lbs.

First Ford tractor with valve-in-head engine tested at Nebraska.

Intercontinental DF Diesel 498
1953-
33.84-31.38 hp @ 1800 rpm, 3 3/4" x 4 1/8";
182 cu in. 4 cyl Buda engine;
4 forward gears, 3.1 to 12.3 mph;
12.41 hp hr/gal; 3591 lbs.

Massey-Harris 44 Special Gasoline 510
1953
48.95-43.58 hp @ 1350 rpm, 4" x 5 1/2";
277 cu in. 4 cyl Continental engine;
5 forward gears, 2.2 to 12.5 mph;
11.06 hp hr/gal; 5789 lbs.

Replaced earlier Model 44 (Test 389) with increased power. Also available in diesel model.

Massey-Harris 33 RT Gasoline 509
1953-1957
39.52-35.54 hp @ 1500 rpm, 3 5/8" x 4 7/8";
201 cu in. 4 cyl Massey-Harris engine;
5 forward gears, 2.8 to 13.5 mph;
10.78 hp hr/gal; 5191 lbs.

Replaced Model 30 (Test 409).

A Model 33 Diesel was also available (Not Tested).

McCormick Super W-4 Gasoline 491
1953-1954
33.85-29.31 hp @ 1650 rpm, 3 1/2" x 4 1/4";
164 cu in. 4 cyl International Harvester engine;
5 forward gears, 2.4 to 15 mph;
10.12 hp hr/gal; 4219 lbs.

1950-1975 FARM TRACTORS

McCormick Super WD-9 Diesel 518
1953-1956
65.19-57.37 hp @ 1500 rpm, 4 1/2" x 5 1/2";
350 cu in. 4 cyl International Harvester engine;
5 forward gears, 2.4 to 15.75 mph;
12.58 hp hr/gal; 9027 lbs.

Replaced Model WD-9 Diesel.

McCormick Farmall Super H Gasoline 492
1953-1954
33.40-30.69 hp @ 1650 rpm, 3 1/2" 4 1/4";
164 cu in. 4 cyl International Harvester engine;
5 forward gears, 2.6 to 16.3 mph;
10.27 hp hr/gal; 4389 lbs.

Replaced Model H Gasoline.

Minneapolis-Moline UB LPG 522
1953-1957
51.27-44.57 hp @ 1300 rpm, 4 1/4" x 5";
283 cu in. 4 cyl Minneapolis-Moline engine;
5 forward gears, 2.5 to 16 mph;
8.15 hp hr/gal; 6359 lbs.

Model UB Gasoline (Test 520) with same displacement and rpm developed 48.38 hp and 10.25 hp hr/gal.

Model U distillate (Test 521) with same displacement and rpm developed 37.23 hp and 8.66 hp hr/gal.

Model U distillate was the last Minneapolis-Moline tractor tested using distillate or tractor fuel.

Willys Farm Jeep Gasoline 502
1953
35.23-27.08 hp @ 2400 rpm, 3 1/8" x 4 3/8";
134.2 cu in. 4 cyl Willys engine;
6 forward gears, 4.6 to 31.3 mph;
8.55 hp hr/gal; 2714 lbs.

4-wheel drive equipped with three-point hitch and hydraulic lift.

FARM TRACTORS 1950-1975

1954

Experiments Pay Off

While results of all laboratory experiments within industry may not be translated into practical, everyday use, they still may have a major impact on the progress of the industry.

Such was the case with the Oliver Corporation's experimental tractor engine, the XO-121. Although this engine never went into production, its design had a marked effect on future tractor power plants. The XO-121 was designed to determine what advantages could be gained in farm tractor engines from higher compression ratios and better fuels. The engine, with a compression ratio of 12 to 1, used a fuel with an octane rating of more than 100. The Ethyl Corporation supplied the fuel for this engine. Compared to Oliver's Model 70, the XO-121 delivered 44 percent more horsepower.

About the same time it was testing the XO-121, Oliver introduced the Super 55, with gasoline or diesel, its first compact utility-type tractor. This was the first Oliver tractor to be equipped with 3-point hitch and draft control. The Oliver Super 66, 77, and 88 replaced earlier models with increased power and displacement.

This planetary type torque amplifier was used on International Harvester tractors to provide 10 forward and two reverse travel speeds. The operator could shift from direct drive to TA or back to direct drive at any time, "on-the-go". Similar "on-the-go" shifting was adopted by other tractor manufacturers in the years to follow.

New Transmission Design

Meanwhile, International Harvester introduced the **torque amplifier**, commonly known as "TA", a major development in transmission design. Located in the clutch housing and connected to the input shaft of the regular 5-speed transmission, the "TA" provided 10 forward and two reverse travel speeds. The operator used a hand lever to shift from direct drive to "TA" or back to direct drive at any time, in any gear-on-the-go.

When in "TA", the travel speed was reduced about 32 percent and the pulling ability increased about 48 percent. This planetary-type "TA" unit was to remain in production on some International Harvester tractors having less than 70 PTO horsepower into 1980. The Super M-TA, Super MD-TA, and McCormick Super W6-TA were the first models equipped with "TA".

Along with the "TA", International Harvester also introduced their first independent or live Power Take-Off. The PTO was driven directly from the engine and was engaged and disengaged by a lever controlling an independent clutch.

Numbers Designate Models

During this period, International Harvester began using the numerical system to designate tractor models. The first tractor using the numerical system was the McCormick Farmall 200, replacing the Super C.

Some manufacturers also were beginning to change their views toward distillate fuel. The year 1954 marked the last year that Minneapolis-Moline produced tractors equipped to use distillate. Gasoline was plentiful, cheap, enabled easier starting, eliminated much oil dilution, and provided more horsepower for the same displacement.

New Upholstered Seat for Massey-Harris Tractors

The new Massey-Harris No. 16 Pacer was introduced with an upholstered seat. Prior to this time, Massey-Harris tractors had used a pressed steel pan seat supported by means of a conical spring.

Minneapolis-Moline placed the Model ZB on the market with new styling and foot-operated clutch. Otherwise, it was similar to the Model Z. A foot-operated clutch was new for Minneapolis-Moline.

The Harris Manufacturing Company

The Harris Manufacturing Company was started in

This is a photograph of Albert W Bonham and the original Power Horse that he developed. This tractor could be controlled from the implement seat by means of pull lines or from a seat located on the rear part of the tractor. (Courtesy Stan Bonham)

1902 by the late George H. Harris in a small workshop in Stockton, California. It was in this workshop that the first Harris combine was built. Within 10 years the founder had enlarged the plant and began the production of the side-hill combine which was to become very popular in the Northwest US.

According to Stan Bonham, his father, Albert Bonham, began building 4-wheel drive tractors in the mid-1930's in Clinton, Utah. Later the tractor was built in Ogden and then in 1939, by the Eimco Corp. in Salt Lake City for about two years. The Bonham Co. continued to build tractors until the war restricted materials. About 1000 units had been built to date. The Power Horse had been using Allis-Chalmers engines. Allis-Chalmers took an interest in Bonham's development and by agreement carried on experimental work for almost another two years. Following a lapse, in 1949 Harris Manufacuring obtained the rights to build the tractor.

The Nebraska tractor test reported the results of three tests made in 1954 (Tests 479, 519 and 523). The three tractors were quite similar except for different types and makes of engines. The tractor was steered by "skidding", much like a crawler tractor. Two control levers pulled back about halfway would stop the tractor. Pulling the levers further back caused the tractor to reverse. If one lever was used the tractor could be turned by releasing the power on that side—further pulling of the lever would put that side in reverse for a sharp turn. Production of the Harris Power Horse discontinued shortly after 1964.

Harris Power Horse 479
1954
33 DB hp @ 2000 rpm, 3 7/16" x 4 1/2";
250.6 cu in. 6 cyl Chrysler engine;
4 forward gears, 2.4 to 15.4 mph;
6.95 hp hr/gal; 5375 lbs.

Massey-Harris No. 16 Pacer Gasoline 531
1954-1957
18.87-17.05 hp @ 1800 rpm, 2 7/8" x 3 1/2";
91 cu in. 4 cyl Continental engine;
3 forward gears, 3 to 7.8 mph;
9.35 hp hr/gal; 2299 lbs.

Upholstered seat.

McCormick Farmall 200 Gasoline   536
1954-1956
24.11-20.92 hp @ 1650 rpm, 3 1/8" x 4";
123 cu in. 4 cyl International Harvester engine;
4 forward gears, 2.5 to 10.6 mph;
8.78 hp hr/gal; 3541 lbs.

Replaced Model Super C.

First International tractor to use a numerical designation.

Oliver Super 55 Gasoline   524
1954-1958
34.39-29.60 hp @ 2000 rpm, 3 1/2" x 3 3/4";
144 cu in. 4 cyl Oliver engine;
6 forward gears, 1.7 to 13.2 mph;
10.08 hp hr/gal; 3369 lbs.

This was the first compact Oliver utility tractor with three-point hitch and draft control.

Model Super 55 Diesel (Test 526) with same displacement and rpm developed 33.71 hp and 12.75 hp hr/gal.

McCormick Farmall Super MTA Gasoline   No Test
(Similar to Test 475)
1954-1955
46 hp @ 1450 rpm, 4" x 5 1/4";
264 cu in. 4 cyl International Harvester engine;
10 forward gears.

This tractor was equipped with Torque Amplifier providing 10 forward travel speeds and partial range power shifting on the go.

First International agricultural tractor equipped with independent PTO.

Oliver Super 77 Diesel   543
1954-1958
44.05-38.13 hp @ 1600 rpm, 3 1/2" x 3 3/4";
216 cu in. 6 cyl Oliver engine;
6 forward gears, 2.5 to 11.5 mph;
13.21 hp hr/gal; 5253 lbs.

Model 77 HC Gasoline (Test 542) with same displacement and rpm developed 43.98 and 9.79 hp hr/gal.

1950-1975 FARM TRACTORS

Oliver Super 88-HC Gasoline 525
1954-1958
55.77-47.08 hp @ 1600 rpm, 3 3/4" x 4";
265 cu in. 6 cyl Oliver engine;
6 forward gears, 2.5 to 12 mph;
10.30 hp hr/gal; 5513 lbs.

Model Super 88 Diesel (Test 527) with same displacement and rpm developed 54.88 hp and 13.41 hp hr/gal.

Model Super 88 LPG tractor was available but not tested.

Oliver XO-121 Gasoline No Test
1954
1600 rpm, 3 3/4" x 4 1/2";
199 cu in. 4 cyl Oliver engine.

Experimental high compression (12 to 1) engine.

Demonstrated at the 1954 Tractor Power and Safety Day, University of Nebraska.

1955

Host of New Models

As farming became more and more specialized, tractor manufacturers began to design models to fit specific needs. The result was a host of models offered by leading companies.

The Ford Motor Company introduced five new tractor models in two power series. Included were row-crop models for use in fields where such crops were grown. Thus, Ford became a multiple-model tractor manufacturer for the first time.

New Utility Tractor

International Harvester brought out its McCormick Farmall 100, 300, 400 and International W-400 series tractors. At the same time, International made available its 300 utility tractor. The utility model had a low profile, making it an ideal tractor for front-end loader work. However, it still maintained most of the features of the Farmall. The torque amplifier (TA) was made available on the 300 series, and a 2-point hitch was offered on both the 300 and 400 series tractors.

High Rating in Fuel Economy

The John Deere 80 replaced the Model R with more power and six forward gears instead of the five in the older model. Tests showed both the R and the 80 to have remarkably good fuel economy ratings.

Efforts continued to improve power output. The Oliver Corporation introduced its Super 99 Diesel, using a 2-cycle engine equipped with a gear-driven blower to assist in scavenging the combustion chamber. Allis-Chalmers installed a gear-driven supercharger on the engine of its Model HD-21 Diesel crawler (Test 550) to increase power output. The next year (1956) Caterpillar D-9 (Test 584) became the first tractor to be tested using a turbocharger.

The new J. I. Case 400 Series tractors were introduced in 1955. The new Series included gasoline, diesel and LPG engines. The new models included an 8-speed transmission, all-gear final drive, and cushioned seats supported by rubber in torsion which could be tilted forward. These features were used in many future Case tractors. This was one of the first of the larger Case tractors to be equipped with a foot operated clutch although a hand clutch was also still available. Power steering, independent PTO, and the Case Eagle Hitch were available. The Series 400 tractors replaced the older Model D tractor that had been introduced in 1939.

Ferguson Improvements

With the introduction of the Ferguson TO-35 gasoline tractor, the foot-operated brake pedals were both mounted on the right-hand side of the tractor. This permitted the operator to use his left foot to control the clutch pedal. This tractor replaced the earlier TO-30.

Another new feature of this new model was the live or continuous PTO. The tractor was operated by a two-stage clutch. When the tractor clutch was depressed in the first stage, the tractor could be stopped without stopping the PTO. By depressing the clutch completely (2nd stage) the PTO could be stopped.

The new "Variable Drive" power take-off was another feature which provided a choice of two different PTO speeds. With a PTO shift lever it is possible to obtain a PTO speed in relation to the ground travel speed of the tractor or in relation to the engine speed of the tractor.

New Minneapolis-Moline Models

About 1955, the Minneapolis-Moline Model ZB, another version of the original Z (Test 438), with the same displacement and rpm, was equipped with disc brakes similar to the Model R.

The Model GB Diesel was announced about 1955 (Test 568). This was the largest horsepower tractor made by the company and was continued in production for about four years. The Lanova system was used in their diesel engines for many years.

1950-1975 FARM TRACTORS

J.I. Case 401 Diesel 565
1955-1957
49.40-43.82 hp @ 1500 rpm, 4'' x 5'';
251 cu in. 4 cyl Case engine;
8 forward gears, 1.4 to 13.7 mph;
13.91 hp hr/gal; 6582 lbs.

Weatherproof cushion on pressed steel seat with rubber in torsion for seat suspension. Seat tilts upward on an angle.

Model 411 Gasoline (Test 566) with same displacement and rpm developed 53.25 hp max and 10.86 hp hr/gal.

Ferguson TO-35 Gasoline 564
1955-1957
33.24-30.51 hp @ 2000 rpm, 3 5/16'' x 3 7/8'';
134 cu in. 4 cyl Continental engine;
6 forward gears, 1.2 to 13.5 mph;
9.93 hp hr/gal; 3093 lbs.

John Deere 80 Diesel 567
1955-1956
65.33-60.04 hp @ 1125 rpm, 6 1/8'' x 8'';
471.5 cu in. 2 cyl John Deere engine;
6 forward gears, 2.5 to 12.25 mph;
15.96 hp hr/gal; 8511 lbs.

Improved version of the Model R.

The starting engine was a 4-cylinder Vee gasoline engine running at 5500 rpm. Bore and stroke 2'' x 2 1/2''.

FARM TRACTORS 1950-1975

Ford 640 Gasoline   560
1955-1957
31.01-28.59 hp @ 2000 rpm, 3.4375" x 3.60";
134 cu in. 4 cyl Ford engine;
4 forward gears, 3.1 to 11.6 mph;
9.87 hp hr/gal; 3031 lbs.

International 300 Utility Gasoline   539
1955-1956
41.26-37.36 hp @ 2000 rpm, 3 9/16" x 4 1/4";
169 cu in. 4 cyl International Harvester engine;
10 forward gears, 1.75 to 16.7 mph;
9.46 hp hr/gal; 4413 lbs.

Model 300 Utility LPG (Test 574) with same displacement and rpm developed 42.68 hp and 8.06 hp hr/gal.

Model 300 Farmall Gasoline (Test 538) with same displacement and 1750 rpm developed 38.16 hp and 9.58 hp hr/gal.

Model 300 Farmall LPG (Test 573) with same displacement and 1750 rpm developed 38.42 hp and 8.16 hp hr/gal.

McCormick Farmall 100 Gasoline   537
1955-1956
20.13-17.83 hp @ 1400 rpm, 3 1/8" x 4";
123 cu in. 4 cyl International Harvester engine;
4 forward gears, 2.3 to 10 mph;
9.06 hp hr/gal; 3038 lbs.

Replaced Model Super A.

McCormick Farmall 400 Gasoline   532
1955-1956
50.78-45.34 hp @ 1450 rpm, 4" x 5 1/4";
264 cu in. 4 cyl International Harvester engine;
10 forward gears, 1.7 to 16.7 mph;
9.53 hp hr/gal; 6519 lbs.

Replaced Model Super MTA.

Model 400 Gasoline was the first tractor tested at Nebraska equipped with the "Torque Amplifier" providing 10 forward travel speeds with partial range operator controlled power-shifting on-the-go.

Model 400 Diesel (Test 534) with same displacement and rpm developed 46.73 hp and 12.31 hp hr/gal.

Model 400 LPG (Test 571) with same displacement and rpm developed 52.36 hp and 8.51 hp hr/gal.

The International W-400 Diesel, Gasoline and LPG models performed essentially the same as the corresponding Farmall models listed above.

1950-1975 FARM TRACTORS

Massey-Harris 555 Gasoline No Test
1955-1958
1350 rpm, 4 1/2" x 6";
382 cu in. 4 cyl Continental engine;
3 to 12 mph; 6920 lbs.

Also available as Model 555 Diesel.

Oliver Super 66 HC Gasoline 541
1954-1958
33.62-27.76 hp @ 2000 rpm, 3 1/2" x 3 3/4";
144 cu in. 4 cyl Oliver engine;
6 forward gears, 2.1 to 10.8 mph;
9.83 hp hr/gal; 3943 lbs.

Replaced the Model 66 HC Gasoline.

Super 66 Diesel (Test 544) same displacement and rpm developed 33.69 hp and 12.65 hp hr/gal.

Minneapolis-Moline GB Diesel 568
1955-1959
62.78-55.44 hp @ 1300 rpm, 4 1/4" x 5";
425.5 cu in. 6 cyl Minneapolis-Moline engine;
5 forward gears, 2.7 to 14.7 mph;
13.44 hp hr/gal; 8170 lbs.

First Minneapolis-Moline Diesel tractor tested at Nebraska.

Model GB Gasoline (Test 547) with 403.2 cu in. displacement and 1300 rpm developed 65.64 hp and 10.10 hp hr/gal.

Model GB LPG (Test 545) with 403.2 cu in. displacement and 1300 rpm developed 70.55 hp and 8.25 hp hr/gal.

1956

More Efficient Deere Tractors

The new model tractors, with their refinements in fuel systems and engine design were setting new records in fuel economy. Tests at the Nebraska laboratory verified this. Three John Deere models, the 520, 620 and 720 established economy records in 10-hour runs. The results were as follows:

Test 590 John Deere 520 LPG 8.87 hp hr/gal.
Test 606 John Deere 720 Distillate 10.03 hp hr/gal.
Test 598 John Deere 620 Gasoline 11.25 hp hr/gal.
Test 594 John Deere 720 Diesel 16.56 hp hr/gal.

This was the last test made at Nebraska on any tractor using distillate. In the same year the Laboratory constructed a new concrete test course, providing a more suitable surface for tests and resulted in more consistant test results.

The John Deere 820 Diesel replaced the John Deere 80 Diesel with increased power in 1956.

The Dubuque built tractors—the John Deere 320 and the 420 were announced in 1956 with increased power. Both tractors were equipped with improved 3-point hitches and hydraulc draft and depth control.

New Massey-Harris Models

The Massey-Harris 333 replaced the earlier Model 33 introduced in 1952 with slightly more power in the gas model. Both tractors were equipped to use gas, diesel or LPG.

A new feature in the Model 333 was the ability to stop the tractor without stopping the PTO. The tractor could be stopped by a hand-lever-operated clutch located in the differential without using the regular foot clutch, and still keep the PTO running.

The Massey-Harris 50 tractor was now available with gas, diesel or LPG. In 1958 this same tractor with sheet metal and color changes was known as the Massey-Ferguson Model 50.

Because of a marketing situation caused by the merger, the Ferguson 40 was announced by the company. This tractor was identical to the 50 Series except for appearance and color.

The Massey-Harris 444 replaced the earlier Model 44. This model had the same arrangement for continuous PTO by means of a clutch in the differential as Model 333.

The Nebraska Tractor Test Course was built in 1956. It is approximately one-half mile for one complete turn. The new Test Course provided much more consistent results than the old earthen one, which was difficult to maintain from one test to another. (Courtesy of University of Nebraska)

1950-1975 FARM TRACTORS

The 444 and the 555 models were the last tractors built in the Racine plant. After 1958 all production was transferred to the Detroit plant.

Drive System Boosts Pull

Minneapolis-Moline introduced Models 445 and 335 which had several new features including independent PTO. Both tractors were equipped with power steering and 3-point hitch with draft control. The Ampli-torc was also intoduced at this time, featuring more travel speeds and easier shifting on-the-go. This permitted operator controlled partial range power shifting and doubled the number of travel speeds. Other new features included a foot-operated clutch and upholstered bucket seat.

The following table gives the results of the benefits of using the Ampli-torc drive to increase the pull. Results were taken from Nebraska Test 624—as an example:

	lbs. Pull	MPH
3rd gear	1599	6.72
3rd gear Ampli-torc	3128	3.34

In Ampli-torc, the pull was about doubled while the travel speed was reduced to one-half of that in direct drive. Some of the later models, like the G 1000 Vista tractors, did not use such a large difference between speed and pull.

Among Other Manufacturers

Allis-Chalmers advertised safety highway lighting, front and rear, as standard equipment on all its WD-45 tractors. Also included was a socket for plugging in a warning light for trailed equipment.

The new 300 Series Case tractors were introduced as the need for row-crop tractors and less horsepower were recognized for smaller farm operations. This new line of tractors was available with gasoline, diesel, distillate, and LPG engines. The 300 Series replaced the VAC models that had been produced since 1942. Transmission options included a standard 4-speed and a 12-speed "Triple-Range". Battery and distributor ignition, available on some earlier models, was now fairly common.

International Harvester added "traction control" to the 2-point hitch on its new line of tractors. With traction control, the operator was able to regulate the amount of weight transfer (front to rear wheels), thus obtaining the best traction and stability for most field applications. The tractor operator could adjust the mechanism by moving a vertical control handle up or down. This same year International Harvester provided hydraulically aided power steering.

One of the more noticeable changes taking place in tractor design during this period was a swing from tricycle-type tractors to the wide front end row-crops. Ford-Ferguson initiated this trend in 1939. In 1956 Ford introduced power steering as standard equipment on all its row-crop tractors.

J.I. Case 301 Diesel   614
1957-1959
30.80-28.73 hp @ 1750 rpm, 3 3/8" x 4 3/8";
157 cu in. 4 cyl Continental engine;
4 forward gears, 2.7 to 12.8 mph;
12.77 hp hr/gal; 3743 lbs.

Model 311 Gasoline (Test 613) with same rpm but smaller displacement of 148 cu in. developed 33.00 hp and 10.07 hp hr/gal.

Cockshutt Deluxe 35 Gasoline   No Test
35 (est) db hp @ 1650 rpm, 3 3/4" x 4 1/2";
198 cu in. Hercules engine;
6 forward gears, 1.9 to 13.4 mph.

FARM TRACTORS 1950-1975

John Deere 320 Gasoline No Test
1956-1958

John Deere 420 Gasoline 599
1956-1958
28.31-26.15 hp @ 1850 rpm, 4 1/4'' x 4'';
113 cu in. 2 cyl John Deere engine;
4 forward gears, 1.6 to 12 mph;
9.72 hp hr/gal; 3581 lbs.

Manfactured at Dubuque.

John Deere 520 Gasoline 597
1956-1958
37.52-33.27 hp @ 1325 rpm, 4 11/16'' x 5 1/2'';
189.8 cu in. 2 cyl John Deere engine;
6 forward gears, 1.5 to 10 mph;
11.10 hp hr/gal; 5809 lbs.

Model 520 All Fuel (Test 592) with same displacement and rpm developed 25.41 hp and 10.03 hp hr/gal.

Model 520 LPG (Test 590) with same displacement and rpm developed 37.24 hp and 8.87 hp hr/gal.

John Deere 620 LPG 591
1956-1958
49.19-44.81 hp @ 1125 rpm, 5 1/2'' x 6 3/8'';
302.9 cu in. 2 cyl John Deere engine;
6 forward gears, 1.5 to 11.5 mph;
8.71 hp hr/gal; 7077 lbs.

Model 620 Gasoline (Test 598) with the same displacement and rpm developed 46.75 hp and 11.25 hp hr/gal.

Model 620 All Fuel (Test 604) with the same displacement and rpm developed 34.57 hp and 9.96 hp hr/gal.

1950-1975 FARM TRACTORS

John Deere 720 Diesel   594
1956-1958

56.66-51.66 hp @ 1125 rpm, 6 1/8" x 6 3/8";
376 cu in. 2 cyl John Deere engine;
6 forward gears, 1.5 to 11.5 mph;
16.56 hp hr/gal; 7899 lbs.

Highest hp hr/gal of any Nebraska tractor test to date.

Model 720 LPG (Test 593) with 360.5 cu in. displacement and same rpm 6" bore developed 57.33 hp and 7.96 hp hr/gal.

Model 720 Gasoline (Test 605) with same displacement and rpm as 720 LPG developed 57.77 hp and 10.93 hp hr/gal.

Model 720 All Fuel (Test 606) with same displacement and rpm as the 720 Gasoline developed 44.13 hp and 10.03 hp hr/gal.

The 730 series tractors were largely a sheet metal change over the 720 series tractors. Test results and performance remained unchanged. The same basic engine used in the 720 series was used in the 730 series.

John Deere 820 Diesel   632
1956-1958

72.82-67.15 hp @ 1125 rpm, 6 1/8" x 8";
471.5 cu in. 2 cyl John Deere engine;
6 forward gears, 2.5 to 12.3 mph;
16.46 hp hr/gal; 8729 lbs.

Replaced John Deere 80 Diesel.

Model 330 Gasoline replaced the 320 Model. Model 330 used the same engine. Power and performance were identical for both tractors. The principle difference was in the sheet metal.

Model 430 replaced Model 420. Basically the engine was the same, the principle difference was in styling sheet metal.

Model 530 series tractors were identical to the 520 series except for sheet metal changes such as improved fenders.

Model 630 series tractors replaced the 620 series. Engines were the same, the principle difference was in sheet metal appearance.

Model 830 Diesel replaced the Model 820. The only real changes were in the appearance.

Ferguson 40 Gasoline   596
1956-1957

32.80-30.84 hp @ 2000 rpm, 3 5/16" x 3 7/8";
134 cu in. 4 cyl Continental engine;
6 forward gears, 1.3 to 14.6 mph;
9.89 hp hr/gal; 3419 lbs.

Similar to Massey-Harris 50 Gasoline except for sheet metal.

International 650 LPG   621
1956-1958

63.91-58.22 hp @ 1500 rpm, 4 1/2" x 5 1/2";
350 cu in. 4 cyl International Harvester engine;
5 forward gears, 2.4 to 15.7 mph;
7.74 hp hr/gal; 8985 lbs.

Replaced International Super WD-9.

Model 650 Gasoline (Test 618) with same displacement and rpm developed 62.11 hp and 9.80 hp hr/gal.

A Model 650 Diesel was also available in 1956. (No Test).

FARM TRACTORS **1950-1975**

Massey-Harris 50 Gasoline   595
1956-1965
32.58-30.50 hp @ 2000 rpm, 3 5/16'' x 3 7/8'';
134 cu in. 4 cyl Continental engine;
6 forward gears, 1.3 to 14.6 mph;
9.74 hp hr/gal; 3432 lbs.

Model 50 LPG (Test 658) with same displacement rpm developed 32.12 hp and 6.86 hp hr/gal.

Model MF 50 Diesel (Test 807) with same rpm and 152.7 cu in. displacement developed 38.33 hp and 13.42 hp hr/gal with Perkins 3-cylinder diesel engine.

Massey-Harris 444 LPG   602
1956-1958
49.24-45.26 hp @ 1500 rpm, 4'' x 5 1/2'';
277 cu in. 4 cyl Continental engine;
10 forward gears, 1.5 to 14.4 mph;
8.32 hp hr/gal; 6385 lbs.

Model 444 Diesel (Test 576) with same displacement and rpm developed 48.21 hp and 13.93 hp hr/gal. Independent PTO.

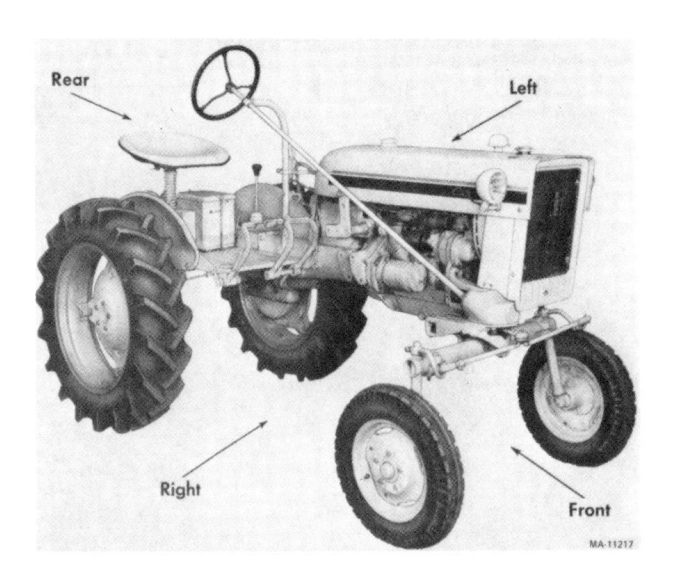

Massey-Harris 333 Diesel   577
1956-1958
37.15-33.34 hp @ 1500 rpm, 3 11/16'' x 4 7/8'';
208 cu in. 4 cyl Continental engine;
10 forward gears, 1.5 to 14 mph;
13.26 hp hr/gal; 6005 lbs.

Replaced 33 series.

Model 333 Gasoline (Test 603) with same displacement and rpm developed 41.89 hp and 10.69 hp hr/gal.

An LPG model was also available (Not Tested).

McCormick Farmall Cub Gasoline   575
1956-1979
10.39-9.87 hp @ 1800 rpm, 2 5/8'' x 2 3/4'';
59.5 cu in. 4 cyl International Harvester engine;
3 forward gears, 2.4 to 7.3 mph;
8.54 hp hr/gal; 1895 lbs.

Replaced the original Cub (Test 386).

1950-1975 FARM TRACTORS

McCormick Farmall 130 Gasoline 617
1956-1958
22.23-19.91 hp @ 1400 rpm, 3 1/8" x 4";
123 cu in. 4 cyl International Harvester engine;
4 forward gears, 2.3 to 10 mph;
10.57 hp hr/gal; 3015 lbs.

Replaced the Model 100.

McCormick Farmall 230 Gasoline 616
1956-1958
28.06-25.00 hp @ 1800 rpm, 3 1/8" x 4";
123 cu in. 4 cyl International Harvester engine;
4 forward gears, 2.7 to 11.7 mph;
10.61 hp hr/gal; 3481 lbs.

Replaced Model 200.

McCormick Farmall 350 Gasoline 611
1956-1958
40.71-37.54 hp @ 1750 rpm, 3 5/8" x 4 1/4";
175 cu in. 4 cyl International Harvester engine;
10 forward gears, 1.7 to 16.1 mph;
10.08 hp hr/gal; 5331 lbs.

Replaced Model 300.

Model 350 LPG (Test 622) with same displacement and rpm developed 41.53 hp and 8.33 hp hr/gal.

Model 350 Diesel (Test 609) with same rpm and 193 cu in. displacement developed 38.65 hp and 12.76 hp hr/gal equipped with Continental Diesel engine.

McCormick Farmall 450 Diesel 608
1956-1958
48.78-45.17 hp @ 1450 rpm, 4 1/8" x 5 1/4";
281 cu in. 4 cyl International Harvester engine;
10 forward gears, 1.7 to 16.6 mph;
12.75 hp hr/gal; 6877 lbs.

Replaced Model 400.

Farmall 450 Gasoline (Test 612) with same displacement and rpm developed 55.28 hp and 10.46 hp hr/gal.

Farmall 450 LPG (Test 620) with same displacement and same rpm developed 54.12 hp and 8.58 hp hr/gal.

FARM TRACTORS 1950-1975

Minneapolis-Moline 335 Gasoline 624
1956-1961
33.50-29.84 hp @ 1600 rpm, 3 3/8" x 4";
165 cu in. 4 cyl Minneapolis-Moline engine;
10 forward gears, 1.44 to 16.5 mph;
9.99 hp hr/gal; 3707 lbs..

1957

Steiger Company Born

On a farm in Northwest Minnesota, two brothers, Douglass and Maurice Steiger, were looking for a larger and more powerful tractor for their farming operations. Unable to find what they needed, they decided to build their own. Using available parts from various sources, they built their first model during the winter of 1957-58.

The Steiger tractor, Model 1, was powered by a 238 hp Detroit Diesel engine. The entire unit weighed 15,000 lbs and was used more than 10,000 hours. The model is now on display in a museum at West Fargo, North Dakota.

Soon, neighbors with the same power needs were ordering tractors and the Steigers were in business. The second generation of tractors consisting of four models, the 1200, 1700, 2200, and 3300, were all built on the Steiger farm. About 120 of these models were to be built and sold during the next several years throughout Saskatchewan, Alberta, Montana and the Dakotas.

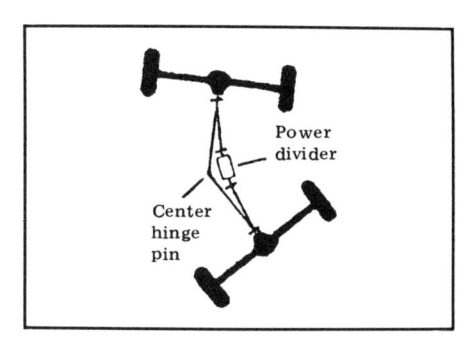

This experimental Ford Typhoon tractor featured a free piston engine, power shift transmission, and sleekly-styled functional lines.

A feature of the Steiger tractors was the Swinging Power Divider which swung to the inside of the turn. This helped reduce the angle of the drive line and improved turning ability of the tractor.

Ford with Turbine Engine

While the Steiger brothers were busy in their farm workshop, the Ford Motor Company was causing considerable excitement with its experimental Typhoon tractor. The Typhoon contained many totally new features. The most spectacular was its unique free piston turbine engine with new power capabilities. The tractor also had a power shift transmission and sleekly-styled functional lines.

Case Merger

In 1957, the J. I. Case Company ventured into more

This free piston turbine engine was used in the Ford Typhoon experimental tractor, developed in 1957.

construction equipment by merging with the American Tractor Corp., a manufacturer of crawler tractors and earth-moving equipment, located at Churubusco, Indiana. This company started building crawler tractors in 1950, manufacturing such models as the Terratrac GT 25 Gasoline, the Terratrac GT 30 Gasoline and Terratrac DT 34 Diesel. This added further diversification and growth for Case. Operations continued at Churubusco until 1961 when production of crawler tractors was transferred to the Case plant at Burlington, Iowa.

1950-1975 FARM TRACTORS

New Transmission Development

A new concept in transmissions for agricultural tractors was introduced by Case. The Case-O-Matic drive for tractors was demonstrated to all dealers at Phoenix, Arizona. New features consisted of a torque converter (replacing the conventional clutch) which could be locked out at will by the operator. When the torque converter was used the tractor had exceptionally good ability to start a heavy load. After gaining normal operation the torque converter could be locked out putting the tractor into direct drive like a conventional clutch. The converter could be re-engaged anytime to provide as much as a 100 percent increase in pull over direct drive to enable the tractor to get through tough spots in the field.

The new Case 600 Series tractors and then the 900 Series, came into being to replace the 500 Series with over 10 percent increase in power largely because of an increase of 150 engine rpm. This new series of tractors was available with diesel and LPG engines. The basic cross-mounted-shaft 4-speed transmission used in the 500 Series was increased to 6-speeds in the 600 and 900. The roller chain final drive and hand clutch, as used in the models L, LA, and 500, was still used.

A "power director" appeared on the new Allis-Chalmers Model D-14. The power director used two clutches—one a conventional foot-operated clutch, the other hand-operated. When the power director was used the forward travel speed was reduced, permitting greater pulling ability. Another innovation of the Model D-14 was its Roll-Shift front axle, which allowed easier spacing of the front wheels when the wide front axle was used.

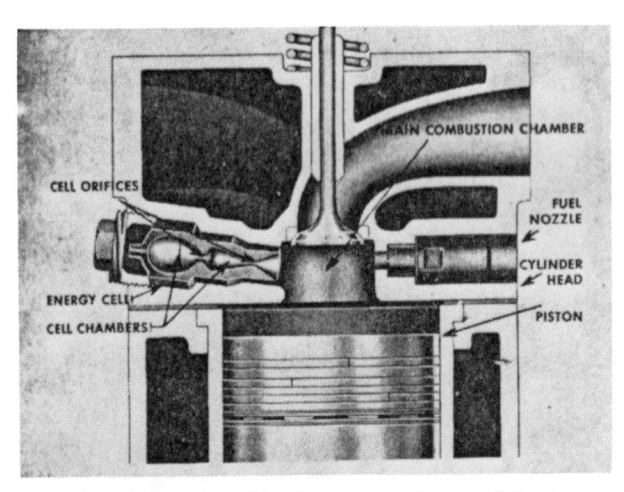

Many of the early Oliver Diesel tractor engines used the Lanova combustion chamber design.

Super Diesel for Oliver

The Oliver Super 99 Diesel replaced the older 1937 Model 99. This model was first built in South Bend, Indiana and later transferred to Charles City, Iowa in 1958, This tractor was available with either a 6-cylinder Oliver Diesel engine or a GM 3-cylinder 2-cycle diesel engine. Both engines were governed to run at 1675 rpm. A torque converter was also available as optional equipment in connection with the regular transmission. An optional cab was available for the Super 99 Model.

The Oliver Super 44 Utility originated at the Battle Creek, Michigan plant and later was built in the South Bend plant. This tractor used a Continental Gasoline engine and was equipped with 3-point hitch and hydraulic control. The unusual feature of this tractor was the off-set location of seat and the engine to provide better forward visibility for the operator doing row-crop work. Production of an improved 440 tractor was later transferred to the Charles City plant.

The Minneapolis-Moline 5-Star replaced the firm's Model UB, with a new appearance and slightly more power, but with the same displacement and rpm for gasoline and LPG. However, the diesel model used a 336 cu in. displacement engine to provide about the same horsepower as the LPG tractor.

The latest David Brown tractor, the 900 model, featured a dual clutch providing live hydraulics and a live PTO.

Allis-Chalmers D-14 Gasoline   623
1957-1960
34.08-30.91 hp @ 1650 rpm, 3 1/2" x 3 7/8";
149 cu in. 4 cyl Allis-Chalmers engine;
8 forward gears, 2 to 12 mph;
10.35 hp hr/gal; 3623 lbs.

Model D-14 LPG (Test 645) with same displacement and rpm developed 31.86 max hp and 7.28 hp hr/gal.

FARM TRACTORS 1950-1975

Allis-Chalmers D-17 Diesel   636
1957-1963
51.14-46.20 hp @ 1650 rpm, 3 9/16" x 4 3/8";
262 cu in. 6 cyl Allis-Chalmers engine;
8 forward gears, 2 to 12 mph;
12.02 hp hr/gal; 4867 lbs.

Model D-17 Gasoline (Test 635) with 4-cylinder 226 cu in. displacement and same rpm developed 52.70 hp and 9.88 hp hr/gal.

Model D-17 LPG (Test 644) with 4-cylinder 226 cu in. displacement and same rpm developed 59.79 hp and 7.73 hp hr/gal.

Case 900-B Diesel   692
1957-1959
70.24-65.59 hp @ 1500 rpm, 4" x 5";
377 cu in. 6 cyl Case engine;
6 forward gears, 2.7 to 13.9 mph;
13.77 hp hr/gal; 8525 lbs.

Replaced Model 500 Diesel (Test 508) which had 63.81 hp at 1350 rpm with same displacement and 14.34 hp hr/gal.

Model 910-B LPG (Test 696) with same displacement developed 71.05 hp at 1350 rpm and 8.11 hp hr/gal.

Cockshutt 540 Gasoline   No Test
No Dates
Manufacturer estimated 31-26 hp @ 1800 rpm, 3 7/16" x 4 3/8";
162 cu in. 4 cyl Continental engine;
6 forward gears, 2.25 to 12 mph;
4295 lbs.

International 330 Utility Gasoline   634
1957-1958
34.24-31.77 hp @ 2000 rpm, 3 1/4" x 4 1/16";
135 cu in. 4 cyl International Harvester engine;
10 forward gears, 1.7 to 16.1 mph;
10.05 hp hr/gal; 4365 lbs.

1950-1975 FARM TRACTORS

Minneapolis-Moline 5 Star LPG 651
1957-1961
54.96-49.36 hp @ 1500 rpm, 4 1/4" x 5";
283 cu in. 4 cyl Minneapolis-Moline engine;
10 forward gears, 1.65 to 17.4 mph;
8.01 hp hr/gal; 6656 lbs.

Replaced Model UB.

Model 5 Star Diesel (Test 652) with 336 cu in. displacement (4 5/8" x 5" bore and stroke) and 1450 rpm developed 54.68 max hp and 12.68 hp hr/gal.

Oliver Super 44 Gasoline No Test
1957-1958
1800 rpm, 3 3/16" x 4 3/8";
139.6 cu in. 4 cyl Continental L Head engine;
4 forward gears, 2.1 to 12.4 mph; 2000 lbs.

Model 440 replaced Model 44 in 1960

Oliver Super 99 Diesel 556
1957-1958
78.74-73.31 hp @ 1675 rpm, 4 1/4" x 5";
213 cu in. 3 cyl GM 2 cycle engine with blower;
6 forward gears, 2.6 to 13.7 mph;
12.38 hp hr/gal; 10,155 lbs.

Someca Som 45 Diesel 625
1957
40.43-37.09 hp @ 1600 rpm, 4.135" x 4.724";
253.5 cu in. 4 cyl O.M. Milano (Italy) engine;
7 forward gears, 1.2 to 13.2 mph;
14.08 hp hr/gal; 5495 lbs.

In 1953, Model DA 50 (Test 495) with 230 cu in. displacement and 1500 rpm developed 38.94 hp and 13.34 hp hr/gal.

Manufactured by Someca 47, BD. Ornano, Saint-Denis-(Seine) France.

Steiger No. 1 No Test
1957-1958
Manufacturer advertised 238 hp @ 2100 rpm, 4 1/4" x 5";
426 cu in. 6 cyl Detroit Diesel engine;
5 forward gears.

First tractor built by Steiger Brothers.

Wagner TR-9 Diesel 631
1957
87.45 db hp @ 1800 rpm, 5 1/8" x 6";
495 cu in. 4 cyl Cummins engine;
10 forward gears, 1.2 to 15 mph;
12.56 hp hr/gal; 15,445 lbs.

Revived interest in 4-wheel drive, replacing many crawler tractors in agriculture.

Manufactured by Wagner Tractor Inc., Portland, Oregon.

Unimog 30 Diesel 607
1957
28.43-25.84 hp @ 2550 rpm, 2.95" x 3.937";
107.8 cu in. 4 cyl Mercedes-Benz engine;
8 forward gears, 0.7 to 33 mph;
11.42 hp hr/gal; 4987 lbs.

First West German unit to be tested at Nebraska.

1950-1975 FARM TRACTORS

1958

Eye Appeal Becomes Important

Styling—Safety—Comfort. These were features which tractor manufacturers were beginning to recognize as important. As the sleek stylish lines of the American automobile gained popularity, so did they gain appeal among potential tractor buyers.

John Deere took these preferences into consideration in producing its 30 Series tractors. These models sported new fenders for the protection of the operator. Sheet metal was altered to give a more pleasing appearance. Dual beam headlights were mounted in the fenders for better visibility at night.

The new Waterloo-built 30 Series tractors included the 530, 630, 730, and 830, the 730 and 830 diesels carried 24-volt electric starters. The new Dubuque-built tractors included the 330 and 430 models.

Diesels Gain Favor

By this time diesel tractor engines were becoming more and more prevalent and popular. Some diesel operators complained about hard starting in cold weather. Various methods have been used to start farm diesel tractor engines. John Deere used a small auxiliary gasoline engine for the John Deere "R" and International used an auxiliary combustion chamber by reducing the compression pressure and operating the engine as a gasoline burner, and then after warm-up, changing to diesel. Some diesels are still using precombustion chambers and glow plugs. Others have used heating elements in the intake system and others have heaters for the crankcase. An ether-base has been used by many injected by a device on the tractor control panel or by a spray can directed at the air intake system. The above methods have all been helpful but it is very important to have a good high capacity battery system and starter which can turn the engine over rapidly.

Case 700 and 800 Series Tractors

The new Case 700 and 800 Series tractors were announced with the 800 Series featuring the Case-O-Matic transmission. The Case 811B (Test 679) Gasoline was the first to use the Case tractor-tested new transmission. This combination of selective gear fixed ratio transmission plus torque converter with lock-out provided 8 forward speeds in direct drive and 8 forward speeds in torque converter drive. The greater selection of travel speeds was a desirable feature for the operator, as well as being typical of the industry trend. The 700 Series was available with a conventional 4-speed gearbox and dual-range, providing 8 forward and 2 reverse speeds.

These models were also equipped with upholstered backrests and cushioned seats supported by rubber springs in torsion. Both 700 and 800 models were available with gasoline, LPG, or diesel 4-cylinder engines.

The new Case 200 Series small tractor with gasoline engine was also announced this year. At 1900 engine rpm, its engine speed was the highest of any Case tractor tested up to this time. Case increased engine speed—the Model LA went to 1150 rpm; the Model 500 reached 1350 in 1953. This trend was evident in the entire tractor industry. The 200 Series was similar to the 300 Series. An 8-speed transmission with synchronized mechanical shuttle, providing 8 forward and 8 reverse speeds, was added to the earlier standard 4-speed and "Tripl-Range" options. A choice of either the Eagle Hitch or 3-point hitch with swinging draft arms was available.

Other 1958 Refinements

Many tractor manufacturers now were coming out with transmission refinements. Massey-Ferguson featured Multi-Power transmission as standard equipment on its MF 180, MF 1100, and MF 1130 tractors, and as a factory-installed option on all other MF tractors except the MF 130. The transmission offered a method of shifting on-the-go to match job requirements. It provided 12 speeds forward and four in reverse. The manufacturer claimed a flip of the Multi-Power lever to

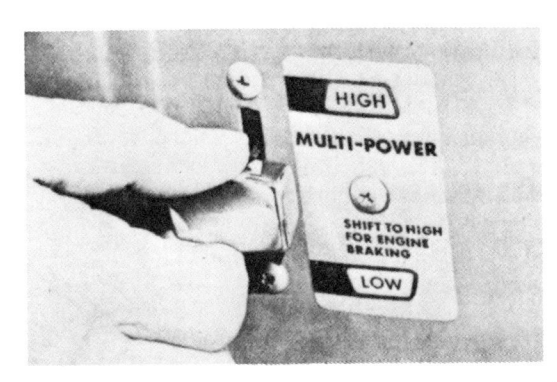

Massey-Ferguson's multi-power transmission offered a method of shifting on-the-go to match job requirements. It doubled travel speeds to 12 forward and four in reverse.

1950-1975 FARM TRACTORS

51

low position would increase the pulling ability by nearly 30 percent while reducing the speed by only 24 percent.

International Harvester initiated a major change in its wheel-type farm tractors with the debut of its Farmall 460 and 560 Series tractors powered by 6-cylinder engines. Rated engine speed for all engines was 1800 rpm. The 460's also were offered in utility models. The 560's, provided in International versions, were commonly referred to as "I" models.

The diesel engines had engine oil-to-engine coolant heat exchangers to cool the crankcase oil, marking the beginning of oil coolers on International Harvester agricultural tractors. The steering shaft was repositioned to run along the engine side rail instead of over the top of the engine. The shaft connected to a worm drive in the front bolster which was hydraulically aided. The hydraulic pump was relocated in the transmission instead of earlier mounting on the engine block.

The MF 65 was a new model with the Multi-Power transmission doubling the number of travel speeds by means of a small lever. This provided operator-controlled power shifting on-the-go currently used on many models. This tractor was available in gas, diesel or LPG models.

In 1958 the name of the company was changed to Massey-Ferguson Inc. from Massey-Harris-Ferguson.

New Models Highlight a Busy Year at Oliver

The Oliver 550 became available with either gasoline or diesel engine as a replacement for the original Super 55 model.

The Oliver 770 Series incorporated a number of improvements over the Super 77, among these were more power, new Power-Booster drive or creeper drive, full-time power steering, new power traction hitch and Power-Juster wheels. Styling was changed; and the tractors painted meadow green with clover white wheels and grille. The 770 Series was available to use gasoline, diesel or LPG. The spurt oil system was discontinued and full oil pressure adopted.

The new Oliver 880 Series replaced the Super 88 tractors with more power, a new Power Booster drive, full-time power steering and a constant-mesh helical gear transmission which replaced the spur-gear transmission. A creeper gear option was also available. This tractor was also made to use gasoline, diesel and LPG.

The Oliver 950, 990 and 995 were the largest Oliver tractors for agricultural use. These tractors replaced the Super 99 Series and were originally manufactured at the South Bend plant; however, production was transferred to the Charles City plant in 1958. The 950 used either an Oliver 6-cylinder gasoline or diesel engine operating at 1800 rpm. The Oliver 990 used a General Motors 3-cylinder, two-cycle, diesel engine operating at 1800 rpm. The Oliver 995 GM Lugmatic diesel was similar to the 990 except it was equipped with a fully automatic torque converter.

David Brown 950 Implematic Diesel 691
1958-1959
39.85-36.50 hp @ 2200 rpm, 3 5/8" x 4";
165 cu in. 4 cyl David Brown engine;
6 forward gears, 1.6 to 10.7 mph;
13.39 hp hr/gal; 4617 lbs.

David Brown's introduction of draft control.

J.I. Case 211-B Gasoline 688
1958-1960
30.84-26.09 hp @ 1900 rpm, 3 1/8" x 4 1/8";
126.5 cu in. 4 cyl Case engine;
4 forward gears, 2.8 to 13.4 mph;
8.88 hp hr/gal; 3655 lbs.

The General Purpose model and the similar 210-B Utility model were also available with a 12-speed "Tripl-Range" transmission.

FARM TRACTORS **1950-1975**

J.I. Case 640-C Gasoline 770
1958-1963
49.72-43.26 hp @ 2250 rpm, 3 13/16" x 4 1/8";
188.4 cu in. 4 cyl Case engine;
6 forward gears, 1.8 to 13.6 mph;
9.17 hp hr/gal; 4675 lbs.

Model 640 Gasoline (Test 773) with same displacement and 2000 rpm developed 50.61 max hp and 9.96 hp hr/gal.

J.I. Case 711-B Gasoline 678
1957-1959
52.19-46.64 hp @ 1500 rpm, 4" x 5";
251 cu in. 4 cyl Case engine;
8 forward gears, 1.4 to 13.7 mph;
9.93 hp hr/gal; 6420 lbs.

Model 701-B Diesel (Test 693) with same rpm but larger engine displacement (267 cu in.) developed 51.20 hp and 13.77 hp hr/gal.

J.I. Case 801-B Diesel 680
1957-1959
54.42-50.14 hp @ 1800 rpm, 4 1/8" x 5";
267 cu in. 4 cyl Case engine;
8 forward gears, 0 to 16.6 mph;
12.13 hp hr/gal; 6935 lbs.

New transmission concept in 1957, Case-O-Matic transmission—selective gear fixed ratio plus torque converter that can be locked out.

Model 811-B Gasoline (Test 679) with 251 cu in. displacement and 1800 rpm developed 53.90 hp and 8.89 hp hr/gal.

Cockshutt 550 Diesel 681
1958
38.45-34.97 hp @ 1650 rpm, 3 3/4" x 4 1/2";
198 cu in. 4 cyl Hercules engine;
6 forward gears, 1.9 to 13.5 mph;
13.42 hp hr/gal; 5695 lbs.

Model 550 Gasoline (Not Tested) with same displacement and rpm. Cockshutt estimated 41.5 max hp.

Cockshutt 560 Diesel 682
1958
48.52-45.61 hp @ 1650 rpm, 4 1/4" x 4 3/4";
269.5 cu in. 4 cyl Perkins engine;
6 forward gears, 1.8 to 13.2 mph;
15.16 hp hr/gal; 7295 lbs.

Cockshutt 570 Diesel 683
1958
60.84-52.25 hp @ 1650 rpm, 3 3/4" x 4 1/2";
298 cu in. 6 cyl Hercules engine;
6 forward gears, 1.8 to 13.2 mph;
13.36 hp hr/gal; 7175 lbs.

Model 570 Gasoline (Not Tested) with same displacement and rpm.

Cockshutt advertised 63.7—53 hp.

Ford 851 Gasoline 640
1958-1962
48.37-41.54 hp @ 2200 rpm, 3.9" x 3.6";
172 cu in. 4 cyl Ford engine;
5 forward gears, 2.3 to 12 mph;
9.25 hp hr/gal; 3445 lbs.

FARM TRACTORS 1950-1975

International 240 Utility Gasoline   668
1958-1962
30.82-28.07 hp @ 2000 rpm, 3 1/8" x 4";
122.7 cu in. 4 cyl International Harvester engine;
4 forward gears, 1.8 to 11.8 mph;
10.07 hp hr/gal; 3637 lbs.

Replaced International 230 Utility.

Farmall 240 Gasoline (Test 667) with same displacement and rpm developed 30.99 hp and 10.23 hp hr/gal.

Massey-Ferguson 50 Diesel   807
1958-1965
38.33-34.20 hp @ 2000 rpm, 3.6" x 5";
152.7 cu in. 3 cyl Perkins engine;
6 forward gears, 1.3 to 14.6 mph;
13.43 hp hr/gal; 3933 lbs.

Model 50 Gasoline (Test 595) with 134 cu in. displacement and same rpm developed 32.58 hp and 9.74 hp hr/gal.

Massey-Ferguson 65 Dieselmatic   808
1958-1965
50.98-45.62 hp @ 2000 rpm, 3.6" x 5";
203.5 cu in. 4 cyl Perkins engine;
12 forward gears, 1.3 to 18.5 mph;
13.9 hp hr/gal; 4511 lbs.

Model 65 Gasoline (Test 659) with 176 cu in. 4-cylinder Continental engine and same rpm developed 46.05 hp and 8.61 hp hr/gal.

Model 65 LPG (Test 657) with 176 cu in. 4-cylinder Continental engine and same rpm developed 42.60 hp and 6.60 hp hr/gal.

McCormick-Farmall 140 Gasoline   666
1958-1979
23.02-21.25 hp @ 1400 rpm, 3 1/8" x 4";
122.7 cu in. 4 cyl International Harvester engine;
4 forward gears, 1.9 to 12.8 mph;
10.61 hp hr/gal; 3000 lbs.

1950-1975 FARM TRACTORS

McCormick-Farmall 340 Gasoline   665
1958-1963
34.74-31.76 hp @ 2000 rpm, 3 1/4" x 4 1/16";
135 cu in. 4 cyl International Harvester engine;
10 forward gears, 1.2 to 16.6 mph;
10.75 hp hr/gal; 4735 lbs.

Replaced Model 330.

Model 340 Diesel (Test 775) with 166 cu in. displacement and same rpm developed 38.93 hp and 11.90 hp hr/gal. Also available as utility model.

The International 330 Utility Gasoline (Test 634) with 135 cu in. displacement and same rpm developed 34.24 hp and 10.05 hp hr/gal.

McCormick-Farmall 460 Gasoline   670
1958-1963
49.47-45.38 hp @ 1800 rpm, 3 9/16" x 3 11/16";
221 cu in. 6 cyl International Harvester engine;
10 forward gears, 1.7 to 16.6 mph;
10.12 hp hr/gal; 5835 lbs.

Operator comfort improved with backrest on seat. First McCormick-Farmall tractor to use a 6-cylinder engine.

Model 460 LPG (Test 676) with same displacement and rpm developed 49.85 hp and 8.08 hp hr/gal. Also available as International 460 Utility LPG.

Model 460 Diesel (Test 672) with 236 cu in. displacement and same rpm developed 50.10 hp and 12.35 hp hr/gal. Also available as International 460 Utility Diesel and International 460 Utility Gasoline.

McCormick Farmall 560 Diesel   669
1958-1963
59.48-54.86 hp @ 1800 rpm, 3 11/16" x 4 25/64";
281 cu in. 6 cyl International Harvester engine;
10 forward gears, 1.5 to 16.6 mph;
13.57 hp hr/gal; 6785 lbs.

First International agricultural tractor to use a 6-cylinder engine. Also available as International 560 Diesel.

Model 560 Gasoline (Test 671) with 263 cu in. displacement (3 9/16 in. bore) and same rpm developed 63.03 hp and 10.73 hp hr/gal.

Model 560 LPG (Test 675) with 263 cu in. displacement and same rpm developed 60.11 hp and 8.55 hp hr/gal.

Oliver 550 Gasoline   697
1958-1975
41.39-35.45 hp @ 2000 rpm, 3 5/8" x 3 3/4";
155 cu in. 4 cyl Oliver engine;
6 forward gears, 1.9 to 14.9 mph;
10.01 hp hr/gal; 3655 lbs.

Replaced Model Super 55.

Model 550 Diesel (Test 698) with same displacement and rpm developed 39.21 and 11.86 hp hr/gal.

FARM TRACTORS **1950-1975**

Oliver 770 Diesel   649
1958-1970
48.80-44.38 hp @ 1750 rpm, 3 1/2" x 3 3/4";
216.5 cu in. 6 cyl Oliver engine;
6 forward gears, 2.1 to 10.8 mph;
12.85 hp hr/gal; 5565 lbs.

Replaced Model Super 77.

Model 770 Gasoline (Test 648) with same displacement and rpm developed 50.04 hp and 10.45 hp hr/gal.

Oliver 880 Diesel   650
1958-1963
59.48-52.64 hp @ 1750 rpm, 3 3/4" x 4";
265 cu in. 6 cyl Oliver engine;
6 forward gears, 2.1 to 10.8 mph;
12.75 hp hr/gal; 5735 lbs.

Replaced Model Super 88.

Model 880 Gasoline (Test 647) with same displacement and rpm developed 61.86 hp and 10.21 hp hr/gal.

Oliver 995 GM Lugmatic Diesel   662
1958-1961
85.37-71.44 hp @ 2000 rpm, 4 1/4" x 5";
213 cu in. 3 cyl GM engine with blower;
12 forward gears, 0.9 to 13.8 mph;
10.19 hp hr/gal; 11,245 lbs.

Model 990 Diesel (Test 661) was similar to the Model 995 except 1800 rpm and no torque converter. It developed 84.10 hp and 12.24 hp hr/gal.

Model 950 Diesel (Test 660) with an Oliver 6-cylinder engine, (4" x 4" bore and stroke), 302 cu in. displacement and 1800 rpm developed 67.23 hp and 12.32 hp hr/gal.

Volvo T-55 Diesel   638
1958
63.86-59.35 hp @ 1800 rpm, 4 1/8" x 5 1/8";
273.6 cu in. 4 cyl Bolinder-Munktell engine;
5 forward gears, 2.5 to 17.2 mph;
15.20 hp hr/gal; 7145 lbs.

Manufacturer A-B Bolinder-Munktell, Eskilstuna, Sweden.

Volvo T-425 Gasoline (Test 637) with 97.6 cu in. displacement and 2000 rpm developed 23.99 hp and 10.01 hp hr/gal.

1959

Standardizing the 3-Point Hitch

With further refinements in 3-point hitches came a need for standardization, so that the farm equipment user could interchange one make of tractor with an implement manufactured by another. This standardization resulted largely through the cooperation of the American Society of Agricultural Engineers (ASAE) and the Farm and Industrial Equipment Institute (FIEI). The action also further established ASAE as a major influence in setting standards for agricultural equipment, thus serving both farmers and the agricultural industry.

In the late 1950's, ASAE, SAE and FIEI proposed standardizing the 3-point free-link attachment for hitching implements to agricultural wheel tractors. ASAE adopted the proposal in March, 1959, subsequently, adopting similar proposals for larger tractors.

The next several years was to see the adoption of two other hitch standardization proposals. One involved tractors equipped with quick-attaching couplers. These standards soon found wide use on nearly all agricultural tractors in the US. The other proposal concerned a 3-point free-link attachment for lawn and garden riding tractors up to 20 horsepower.

Case Speeds Up

The new Case 400 Series tractor made its debut in 1959. This series was built at the Rock Island plant, and was similar in design to the 200 and 300 Series of tractors introduced earlier. The Case-O-Matic transmission was featured and could be combined with the 8-speed shuttle, 8-speed dual range, or standard 4-speed options. This was an entirely different model from the 400 tractor introduced in 1955. Rated speed was 2000 rpm, the highest of any Case agricultural tractor at this time.

Massey Acquires Perkins

Massey-Ferguson acquired F. Perkins Ltd. of Peterborough, England, providing a good selection of engines. From this time on nearly all Massey-Ferguson tractors were equipped with Perkins engines.

The Massey-Ferguson 35 Diesel was announced. It was very similar to earlier models of the MF 35 except a Perkins 3-cylinder diesel engine was used.

The MF 85 was announced in 1959 and was available in gasoline, diesel and LPG models. This was the last tractor built by Massey-Ferguson to use LPG. LPG was losing favor among users due to the increase in price and the extra cost and inconvenience of the fuel.

The Massey-Ferguson Super 90 soon replaced the Model 85. The gasoline models were very similar; however, the Massey-Ferguson Super 90 used a Perkins diesel engine instead of the Continental engine used in the Model 85. The Massey-Ferguson Super 90 was also available with either an 8-speed transmission or a 16-speed shift-on-the-go transmission.

This controller regulated the speed and reversed the Allis-Chalmers experimental fuel cell tractor. The power generated went through bus bars to the controller which placed the four banks of fuel cells in series or parallel, and thus varied the amount of electricity reaching the motor. The reversing lever changed the polarity of the current reaching the motor.

Electric Tractor Introduced

A completely new approach to farm tractor power appeared in 1959 with Allis-Chalmers' fuel cell tractor. Strictly experimental, it set scientists and engineers pondering the practicality of "electric" tractors.

For the fuel cell research project, Allis-Chalmers used its Model D-12 tractor chassis. The electricity driving the tractor came from 1008 individual fuel cells. A mixture

1950-1975 FARM TRACTORS

of gasses, mostly propane, fueled the individual cells. The chemical reaction within the cells created a direct current flow. This flow was then used to operate the tractor which developed about 20 horsepower. The firm's engineers felt their fuel cell power engine held exciting promise. They noted that it offered twice the efficiency of other engines of the period, had no moving parts, no exhausts or odorous fumes, and "ran without a whisper."

Following are some statistics on the fuel cell tractor:
  Fuel cell dimensions—one quarter inch thick, 12 inches square.
  Fuel cell voltage—about one volt open circuit.
  Total electrical output—15 kw.
  Fuel gas used—a mixture, largely propane.
  Type of motor—standard Allis-Chalmers 20 hp dc.
  Drawbar pull—3000 lbs.
  Tractor weight—5270 lbs.

The tractor was never marketed, but the lessons which scientists and engineers gained from it were destined to have impact on tractor power sources for years to come. The tractor now is in the Smithsonian Institution in Washington, DC.

Allis-Chalmers experimental fuel cell tractor represented a new approach in farm tractor power. The electricity driving the tractor came from 1008 individual fuel cells. The tractor developed about 20 horsepower.

Debut for Porsche Tractor

At the time Allis-Chalmers scientists were working on their experimental fuel cell tractor, their West German colleagues in Friedrichshafen were building an air-cooled diesel for the Porsche tractor. Four different sizes were made available in the US—a 1-cylinder, a 2-cylinder, a 3-cylinder, and a 4-cylinder. Two of the Porsche diesels were submitted to the Nebraska Lab for official testing in 1959 (Test 699 and 728). Several small garden tractors using air-cooled engines had been tested at the laboratory, but Porsche was the first to have a farm tractor equipped with an air-cooled system in the power plant. A unique feature of the Porsche tractor cooling system was the gear-driven blower fan. The system used no belts.

Another feature of the Porsche tractors was the ease in which the operator could raise the hood to service the engine. He could lift the entire hood over the engine by unhooking two latches on the side of the tractor.

In the early 1960's, Renault acquired manufacturing rights of the Porsche tractor and assumed responsibility for future service.

Also abroad, the International Harvester Company of Great Britain, was producing the B-275, one of the first International tractors to be imported into the US. It was powered by a 4-cylinder diesel engine having a displacement of 144 cu in. which operated at 1875 rpm, and produced 32.88 PTO horsepower. The transmission had four speeds, followed by a Hi-Lo range transmission which provided eight forward and two reverse speeds. A differential lock, actuated by a foot pedal, was standard on these tractors which were among the first Internationals so equipped.

In the US, International Harvester was taking steps to increase its emphasis on research and engineering as it opened its new Farm Equipment Research and Engineering Center south of Hinsdale, Illinois. The new complex, comprising 442 acres, was claimed to be the largest of its kind in the world.

New designs in transmission systems continued. Ford announced its Select-O-Speed transmission. It provided 10 forward travel speeds with a power shift unit controlled by the operator by means of a small lever located below the steering wheel. No clutch was used; it was replaced by a small pedal intended to release the hydraulic oil pressure, if necessary.

Engineers had on their drawing boards various plans for large, 4-wheel drive tractors. The first John Deere 4-wheel drive was the 8010 Diesel. One year later this model was replaced with the John Deere 8020 Diesel. It was essentially the same as the 8010, with some improvements in clutch and transmission design. Deere built only a limited number of these tractors due to market conditions.

Minneapolis-Moline replaced the Model 445 with the 4 Star which was available with 206 cu in. gasoline, LPG or diesel engines. The diesel version of the 4 Star was produced in very limited numbers. The 4 Star was essentially the same as the 445 except in appearance and an improved hydraulic system. Other model changes saw the Model 335 replaced by the Jet Star tractor, and the introduction of the G-VI, to replace the GB.

The Oliver 660 Series replaced the earlier Super 66 tractor and was available with either a 4-cylinder gasoline or diesel engine at 2000 rpm.

FARM TRACTORS 1950-1975

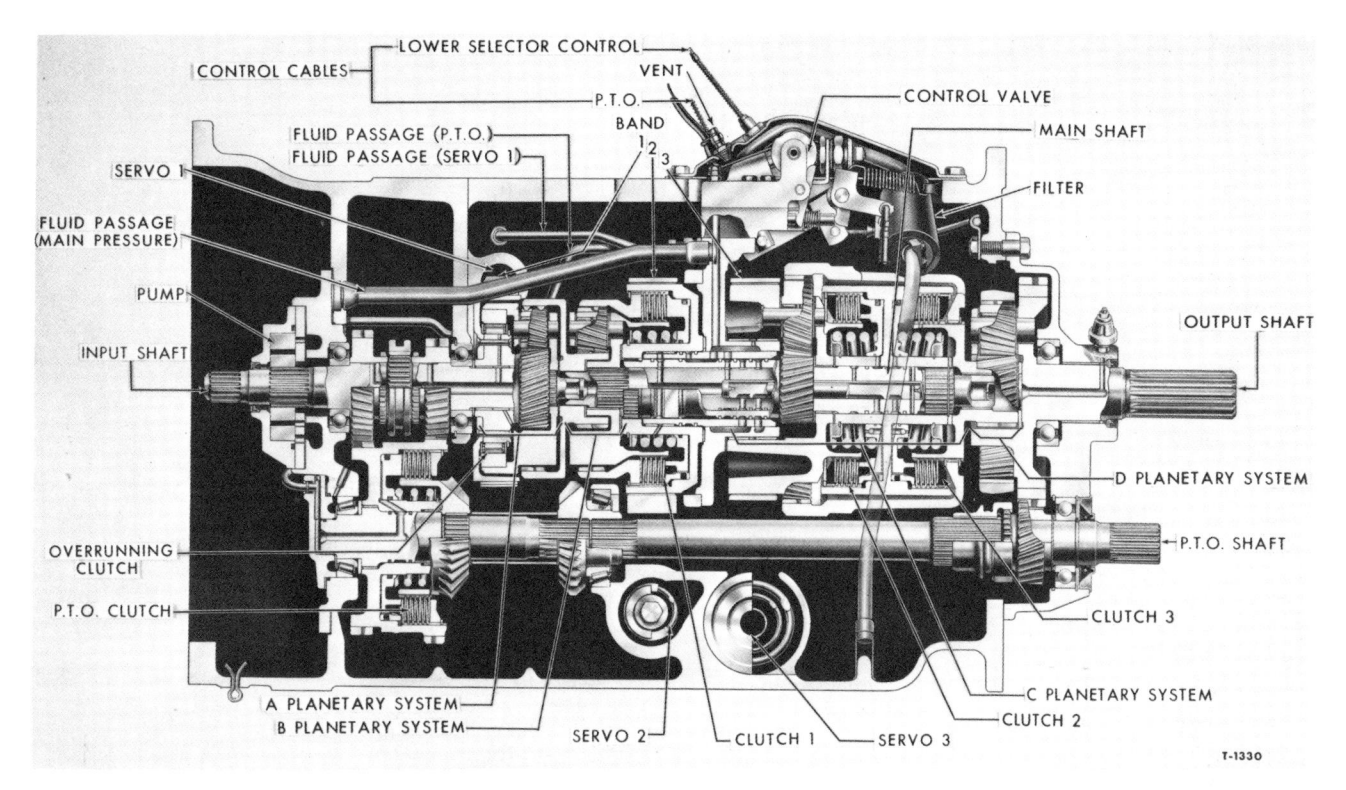

Cross section of Select-O-Speed transmission introduced by Ford in 1959. This made possible operator-controlled full power shift in all speed ranges by means of a small control lever under steering wheel. A foot pedal was available to release hydraulic oil pressure if needed.

Allis-Chalmers D-10 Gasoline   724
1959-1964
28.51-25.84 hp @ 1650 rpm, 3 3/8" x 3 7/8";
138.7 cu in. 4 cyl Allis-Chalmers engine;
4 forward gears, 2.0 to 11.4 mph;
10.77 hp hr/gal; 2860 lbs.

Allis-Chalmers D-12 Gasoline   723
1959-1964
28.56-24.86 @ 1650 rpm, 3 3/8" x 3 7/8";
138.7 cu in. 4 cyl Allis-Chalmers engine;
4 forward gears, 2.0 to 11.4 mph;
10.28 hp per hr/gal; 2945 lbs.

1950-1975 FARM TRACTORS

J.I. Case 411-B Gasoline   689
1959-1960
37.12-30.67 hp @ 2000 rpm, 3 3/8" x 4 1/8";
148 cu in. 4 cyl Case engine;
8 forward gears, 0 to 11.6 mph;
7.95 hp hr/gal; 4349 lbs.

Selective gear fixed ratio plus torque converter with lock out.

Series 400 tractors were built at Case's Rock Island (IL) plant.

This model was also available with LPG engine.

J.I. Case 831-C Diesel   736
1959-1969
63.74-58.48 hp @ 1900 rpm, 4 3/8" x 5";
301 cu in. 4 cyl Case engine;
8 forward gears, 0 to 17.6 mph;
13.20 hp hr/gal; 7275 lbs.

Model 831 Diesel (Test 737) with same displacement and 1700 rpm developed 63.76 max hp and 13.61 hp hr/gal.

Model 841 Gasoline (Test 738) with 284 cu in. displacement and 1700 rpm developed 64.57 max hp and 10.19 hp hr/gal.

Model 841 LPG (Test 740) with 284 cu in. displacement and 1700 rpm developed 63.38 hp and 7.94 hp hr/gal.

J.I. Case 930 Diesel   741
1959-1969
80.65-71.92 hp @ 1600 rpm, 4 1/8 x 5";
401 cu in. 6 cyl Case engine;
6 forward gears, 2.5 to 13.6 mph;
13.55 hp hr/gal; 8845 lbs.

Replaced Model 900-B (Test 692) with over 10 horsepower increase as a result of larger displacement and higher engine rpm.

The chain final drive was discontinued in 1969.

Model 940 LPG (Test 739) with 377 cu in. displacement and same rpm and 4" x 5" bore and stroke developed 79.64 max hp and 7.54 hp hr/gal.

John Deere 435 Diesel   716
1959-1960
32.91-28.41 hp @ 1850 rpm, 3 7/8" x 4 1/2";
106.1 cu in. 2 cyl GM engine with blower;
5 forward gears, 1.9 to 13.5 mph;
11.61 hp hr/gal; 4101 lbs.

62

FARM TRACTORS **1950-1975**

John Deere 8010 Diesel No Test
1959-1964
Mfg. est. 215 hp @ 2100 rpm, 4.25" x 5.00";
425 cu in. 6 cyl GM engine;
9 forward gears, 2 to 18 mph; 20,700 lbs.

Model 8020 was identical to Model 8010 except for improved clutch and transmission.

Fiat 411-R Diesel 707
1959
36.75-33.24 hp @ 2300 rpm, 3.35" x 3.94";
138.5 cu in. 4 cyl Fiat engine;
6 forward gears, 1.4 to 14.2 mph;
12.16 hp hr/gal; 3565 lbs.

Manufactured by Fiat S.p.A., Turin, Italy.

First tractor tested at Nebraska using radial tires (Pirelli).

Ford 681 Diesel 706
1957-1961
31.56 @ 2200; 25.61 @ 2000 rpm, 3.562" x 3.6";
144 cu in. 4 cyl Ford engine;
10 forward gears, 1.0 to 16.9 mph;
12.10 hp hr/gal; 3490 lbs.

Ford Select-O-Speed transmission.

Ford 881 Gasoline 701
1957-1961
46.16 @ 2200; 37.07 @ 2000 rpm, 3.9" x 3.6";
172 cu in. 4 cyl Ford engine;
10 forward gears, 1.0 to 16.4 mph;
8.94 hp hr/gal; 3563 lbs.

Introduction of Ford Select-O-Speed transmission providing operator controlled full range power-shifting.

Ford 881 Diesel 705
1957-1961
41.36 @ 2200; 33.91 @ 2000 rpm, 3.9" x 3.6";
172 cu in. 4 cyl Ford engine;
10 forward gears, 1.0 to 16.4 mph;
12.86 hp hr/gal; 3635 lbs.

1950-1975 FARM TRACTORS

Ford 841 Diesel   653
1957-1961
39.88-36.77 hp @ 2000 rpm, 3.9'' x 3.6'';
172 cu in. 4 cyl Ford engine;
12 forward gears, 2.5 to 15.5 mph;
13.49 hp hr/gal; 3559 lbs.

First US built Ford Diesel tractor engine.

International B-275 Diesel   733
1959-1964
32.88-30.76 hp @ 1875 rpm, 3 3/8'' x 4'';
144 cu in. 4 cyl International Harvester engine;
8 forward gears, 1.6 to 14 mph;
11.81 hp hr/gal; 3736 lbs.

Built by International Harvester in the United Kingdom.

Fordson Dexta Diesel   684
1958-1961
31.41-27.57 hp @ 2000 rpm, 3.5'' x 5'';
144 cu in. 3 cyl Ford English engine;
6 forward gears, 1.6 to 17.3 mph;
12.54 hp hr/gal; 3393 lbs.

International 660 Diesel   715
1959-1963
78.78-71.38 hp @ 2400 rpm, 3 11/16'' x 4 25/64'';
281 cu in. 6 cyl International Harvester engine;
10 forward gears, 2 to 15.4 mph;
12.72 hp hr/gal; 9875 lbs.

First use of planetary drive on outer axle on International agricultural tractor.

Replaced Model 650 Diesel.

Model 660 Gasoline (Test 721) with 263 cu in. displacement and same rpm developed 81.39 hp and 10.33 hp hr/gal.

Model 660 LPG (Test 722) with same displacement and rpm as Model 660 Gasoline developed 80.63 hp and 7.55 hp hr/gal.

FARM TRACTORS **1950-1975**

Massey-Ferguson 35 Diesel 744
1959-1965
37.04-33.02 hp @ 2000 rpm, 3.6'' x 5'';
152.7 cu in. 3 cyl Perkins engine;
6 forward gears, 1.3 to 14.6 mph;
13.88 hp hr/gal; 3559 lbs.

Similar to Model TO-35 Diesel (Test 690) except for a Perkins engine.

Massey-Ferguson 85 LPG 727
1959-1962
62.21-56.47 hp @ 2000 rpm, 3 7/8'' x 5 1/8'';
242 cu in. 4 cyl Continental engine;
8 forward gears, 1.7 to 21.5 mph;
7.30 hp hr/gal; 5753 lbs.

Model 85 Gasoline (Test 726) with same displacement and rpm developed 61.23 and 9.48 hp hr/gal.

Minneapolis-Moline 4 Star Gasoline 789
1959-1962
44.57-39.55 hp @ 1750 rpm, 3 5/8'' x 5'';
206.5 cu in. 4 cyl Minneapolis-Moline engine;
10 forward gears, 1.7 to 17.8 mph;
9.65 hp hr/gal; 5245 lbs.

Replaced Model 445.

Model 4 Star LPG (Test 790) same displacement and rpm developed 45.54 hp and 7.23 hp hr/gal.

Minneapolis-Moline GVI LPG 791
1959-1962
78.44-71.15 hp @ 1500 rpm, 4 1/4'' x 5'';
425.5 cu in. 6 cyl Minneapolis-Moline engine;
5 forward gears, 3.1 to 17.1 mph;
7.64 hp hr/gal; 8155 lbs.

Replaced Model GB.

Model GVI Diesel (Test 792) with same displacement and rpm developed 78.49 hp and 12.25 hp hr/gal.

Oliver 660 Gasoline   No Test
1959-1964
2000 rpm, 3 5/8" x 3 3/4";
155 cu in. 4 cyl;
6 forward gears, 2.3 to 6.9 mph; 3378 lbs.

Replaced Super 66 models.

Diesel version also produced.

Oliver Twin 880 Gasoline   No Test
1959
This was an experimental combination of tractors used for parades, shows, and for publicity purposes.

Porsche L318 Super Diesel   728
1959
37.21-33.40 hp @ 2300 rpm, 3.74" x 4.56";
150.6 cu in. 3 cyl Porsche air-cooled engine;
8 forward gears, 0.7 to 12.3 mph;
10.88 hp hr/gal; 4839 lbs.

Manufactured by Porsche-Diesel Friedrichshafen, West Germany.

Porsche Junior L108 Diesel   699
1959
11.29-9.58 hp @ 2250 rpm, 3.74" x 4.56";
50.2 cu in. 1 cyl Porsche air-cooled engine;
6 forward gears, 1.2 to 12.9 mph;
10.73 hp hr/gal; 2575 lbs.

Manufactured by Porsche-Diesel Friedrichshafen, West Germany.

FARM TRACTORS 1950-1975

Wagner TR-14A Diesel 700
1959
155.25 drawbar hp @ 2100 rpm, 5 1/8'' x 6'';
743 cu in. 6 cyl Cummins engine;
10 forward gears, 2.1 to 20.9 mph;
12.89 hp hr/gal; 21,225 lbs.

1960

Tandem Hitching

They called it "tandem hitching"—connecting two tractors together to get more drawbar power. Farmers and engineers in many areas had tried tandem hitching at one time or another, with varying degrees of success. Some suppliers offered kits to farmers and dealers to make these conversions. Researchers at Iowa State University developed a tandem hitch which involved removing the front wheels of both tractors, hitching them together with a vertical hinge, and providing suitable controls for the operator.

Semi-tandem hitch probably was more common. It required the removal of the front wheels of the rear tractor only. A long pull bar was attached to the rear tractor and pinned to the drawbar of the front tractor. The operator controlled both tractors with improvised controls from the front tractor.

Ford experimented with a semi-tandem hitch designed to let the operator ride on the rear tractor, but the hitch saw only limited production.

End of 2-Cylinder John Deere

The era when the pop-pop-pops of those 2-cylinder John Deeres echoed through the crisp morning air of many a countryside was now destined to end. The 2-cylinder tractor had been in use for almost half a century. Now the demand for more power made it mandatory to develop multi-cylinder engines to satisfy desires for more power and modern requirements.

Ease, Comfort & Visibility

John Deere referred to its new 4-cylinder tractors as

With this semi-tandem hitch, the operator could control both tractors from the rear vehicle. This Ford hitch arrangement saw only limited production. The clock at right indicates time required for one man to install.

1950-1975 FARM TRACTORS

the "New Generation." These new models were the 1010, 2010, 3010, and 4010. The new line had more travel speeds, a larger selection of engine speeds, significantly more power than their predecessors. The 3010 and 4010 were equipped with larger fuel tanks which were mounted in front of the radiator. Both these tractors, as well as the 2010, were available as gasoline, diesel, or LPG users.

The 1010, 2010, and 3010 tractors were equipped with 4-cylinder engines. The 4010 had six cylinders. All the new tractors came with 3-point hitches with hydraulic load and depth control. Power brakes were standard on the 3010 and 4010 tractors. The two larger tractors also used hydrostatic power steering. This system disconnected the steering wheel (mechanically) from the front wheels. All steering was controlled hydraulically. The ease of operation and the comfort seat indicated that John Deere had considered human engineering in their new designs.

504 diesel, with 187.5 cu in. displacement, operated at 2200 rpm.

White Enters Tractor Field

The Oliver 500 gasoline or diesel was manufactured especially for Oliver by David Brown of England. The Model 500 was originally distributed from the Cleveland plant until it was transferred to Charles City in 1961. Regular equipment included 3-point hitch, hydraulic system, differential lock, PTO, traction control system and lights.

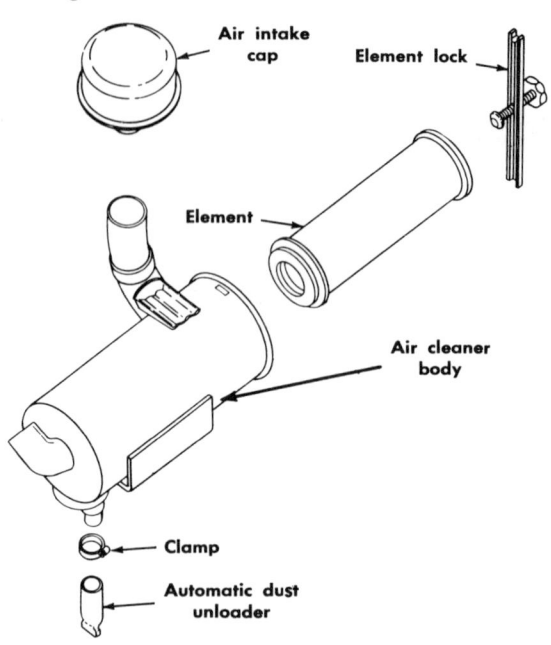

Dry-type air cleaner on International 504 diesel tractors was easy to service. Drawings show air cleaner components.

Mark Opening of Decade

The International Harvester Company marked the opening of the 1960 decade by announcing four major design advancements for its new 404 and 504 model tractors. Those advancements were:

1. Their first US designed 3-point hitch with hydraulic draft control. The top link draft sensing was accomplished by using a torsion bar. This was unique for draft control systems on the market at this time.

2. An oil cooler to cool the transmission and hydraulic oil. As all tractor manufacturing firms increased horsepower, more loads were imposed on all working parts of the tractors. Thus, it became necessary to provide cooling systems not only for the engine crankcase but also for the transmission and hydraulic oil systems.

3. A dry-type air cleaner. Research work was being carried out by many tractor manufacturers and air cleaner companies to develop a dry-type air cleaner that would do a better cleaning job and be easier to service.

4. Hydrostatic power steering on the International 504 model, thus eliminating any mechanical connection between the steering wheel and the wheels to be steered.

The 404 and 504 models were offered in both Farmall and International versions. The 404 was available with engines burning gasoline or LPG, while the 504 was provided with gasoline, LPG or diesel-fueled engines. The engines in the 404's had a 134.8 cu in. displacement operating at 2000 rpm. The engines in the 504 gasoline and LPG models had a 152.1 cu in. displacement. The

The Oliver 440 tractor was a continuation of the Super 44 tractor. It used a 4-cylinder Continental Gasoline engine with an "L" head valve arrangement, 3-3/16 in. × 4-3/8 in. bore and stroke operating at 1800 rpm. This one-row type of tractor had the seat and the engine located off-side providing better visibility for the operator.

Three different series, "A", "B", and "C", of the Oliver Model 1800 tractors were manufactured during this period. Series "A" 1960 to 1962; Series "B" 1962 to 1963; and Series "C" 1963 to 1964. A choice of 6-cylinder gasoline, LPG or diesel engines were available.

The Oliver Model 1800 "A" Diesel Row Crop was a 1960 anniversary model commemorating 25 years of Oliver experience manufacturing 6-cylinder engines. The engines were basically the same except the diesel had 283 cu in. displacement while the gasoline had 265 cu in. displacement. The Model 1800 "C" operated at 2200 engine rpm; the 1800 "B" operated at 2000 rpm.

The Oliver 1900 Series was also built as "A", "B", and "C" models. These three models used General Motors 4-cylinder 2-cycle engines. The 1900 "A" used a GM series 4-53 Model 5121 engine, the 1900 "B" and "C" used the Model 5221 in same series engine. Some had a governed engine speed of 2000 and others 2200 rpm. The 1900 series had the same features as the 1800 and were also available with front-wheel drive.

Massey-Ferguson purchased 500 Model 98 Diesel tractors from Oliver in 1959 and 1960. These tractors were basically the same as the Oliver 990 except for sheet metal, paint and decals.

On November 1, the White Motor Co., of Cleveland, a builder of heavy-duty trucks, acquired Oliver as a wholly-owned subsidiary. Oliver's manufacturing plants were then consolidated into three main plants which were then located at Charles City (tractors); South Bend (plows and tillage tools); and at Shelbyville, Illinois (haying equipment).

New Series for Case

A new series of tractors, the 430, 530, 630, 730, 830 and 930, referred to as the "30" Series tractors, was developed by Case. These tractors used a different model designation than had been used before. The second digit indicated the fuel use. For example (3) indicated diesel and (4) indicated a spark ignition engine using either gasoline or LPG fuels. Distillate had been discontinued on all but a few small Case tractors in 1954.

The new models replaced the older Series 400, 500, 600, 700, 800 and 900 Model tractors. Horsepower had been increased on many models, mainly by higher engine rpm or larger displacement or both. The Model 640 C Gasoline had a rated engine speed of 2250 rpm—the highest of any Case tractor engine up to that time.

The 430, 530, and 630 models, built at Rock Island, were the "small" tractors. They were similar in design to the earlier 200, 300, 400, 500 and 600 Series tractors, but offered more refinements and options, and improved performance. The 730 and 830 models were both available with either Case-O-Matic or standard dry clutch 8-speed transmissions, and a choice of gasoline, LPG, or diesel engines.

A 1000 rpm PTO option was offered for the first time on the 930 Series, in addition to the 540 rpm option. The oil-lubricated hand clutch, 6-speed transmission, and chain final drive were continued from the 900 Series.

New Czech Tractor Tested

At Nebraska, a Czechoslovakian tractor, the Zetor 50 Super, was tested. This tractor was equipped with an air compressor, hydraulic brakes, and an unusual oil cleaner. The oil cleaner operated as a centrifugal spinner, driven by the pressure of the oil entering the filter or chamber. The operation was similar to that of the bowls in those old cream separators found on many American farms in the early 1900's.

From England, David Brown, Ltd. introduced an improved version of its 950 diesel. This tractor was equipped with a multi-speed PTO providing 540 or 1000 rpm. These PTO speeds were obtained at 1828 or 2000 engine rpm, respectively.

The Minneapolis-Moline M-5 Series was introduced to replace the earlier 5 Star. The model was in production from 1960 to 1963. This series was available in gasoline, LPG and diesel versions. The model was also known as the M-504 and had a front axle drive.

The Holder four-wheel drive tractors, manufactured by the Gebrüder Holder Maschinenfabrik, West Germany had been marketed by Tradewinds, Inc. Tacoma, Washington for many years and continued until late 1970. These diesel tractors ranged in size from 10 to 48 advertised horsepower.

Allis-Chalmers D-15 Gasoline 795
1960-1962
40.00-35.94 hp @ 2000 rpm, 3 1/2" x 3 7/8";
149 cu in. 4 cyl Allis-Chalmers engine;
8 forward gears, 1.9 to 15.3 mph;
9.81 hp hr/gal; 3985 lbs.

Model D-15 Diesel (Test 796) with 175 cu in. displacement and same rpm developed 36.51 hp and 9.86 hp hr/gal.

Model D-15 LPG (Test 797) with same displacement and rpm as the gasoline model developed 37.44 max hp and 7.81 hp hr/gal.

Allis-Chalmers (Series II) D-15 Gasoline 837
1960-1968
46.18-39.60 hp @ 2000 rpm, 3 5/8" x 3 7/8";
160 cu in. 4 cyl Allis-Chalmers engine;
8 forward gears, 1.8 to 15.3 mph;
10.16 hp hr/gal; 4025 lbs.

Model (Series II) D-15 LPG (Test 838) is very similar to the gasoline model except the horsepower is 43.55 max and 7.91 hp hr/gal.

David Brown 850 Diesel   734
1960-1965
33.56-31.84 hp @ 2000 rpm, 3 1/2" x 4";
154 cu in. 4 cyl David Brown engine;
6 forward gears, 2.2 to 14.7 mph;
13.65 hp hr/gal; 4179 lbs.

Model 850 Gasoline (Test 735) same displacement and rpm developed 32.00 hp and 8.30 hp hr/gal.

J.I. Case 441 Gasoline   774
1960-1969
33.11-29.39 hp @ 1750 rpm, 3 3/8" x 4 1/8";
148 cu in. 4 cyl Case engine;
4 forward gears, 2.5 to 11.9 mph;
10.36 hp hr/gal; 3620 lbs.

Also J. I. Case 470 Gasoline. Replaced Rock Island built 400 Series tractors.

J.I. Case 431 Diesel   785
1960-1969
34.38-31.33 hp @ 1750 rpm, 3 13/16" x 4 1/8";
188.4 cu in. 4 cyl Case engine;
4 forward gears, 2.5 to 11.9 mph;
14.16 hp hr/gal; 3839 lbs.

J.I. Case 531 Diesel   772
1960-1969
41.27-37.20 hp @ 1900 rpm, 3 13/16" x 4 1/8";
188.4 cu in. 4 cyl Case engine;
12 forward gears, 1.6 to 22.4 mph;
14.71 hp hr/gal; 4117 lbs.

Model 531-C Diesel (Test 786) same displacement and 2100 rpm developed 40.25 hp and 11.74 hp hr/gal.

Also Case 570 Diesel.

FARM TRACTORS 1950-1975

J.I. Case 541-C Gasoline　　　769
1960-1969
41.26-37.14 hp @ 2100 rpm, 3 1/2" x 4 1/8";
158.7 cu in. 4 cyl Case engine;
8 forward gears, 0 to 14.8 mph;
9.14 hp hr/gal; 4345 lbs.

Model 541 Gasoline (Test 771) with same displacement and 1900 rpm developed 39.50 hp and 9.76 hp hr/gal (also Model 570).

Model 531 Diesel (Test 772) with 188.4 cu in. displacement and 1900 rpm developed 41.27 hp and 14.71 hp hr/gal (also Model 570).

J.I. Case 731-C Diesel　　　742
1960-1969
56.50-50.64 hp @ 1900 rpm, 4 1/8" x 5";
267 cu in. 4 cyl Case engine;
8 forward gears, 1.7 to 17.6 mph;
11.45 hp hr/gal; 7095 lbs.

Model 741-C Gasoline (Test 778) with 4" x 5" bore and stroke (251 cu in. engine) and 1900 rpm developed 57.88 hp and 8.92 hp hr/gal.

Model 741-C LPG (Test 781) with 251 cu in. displacement and 1900 rpm developed 57.19 hp and 6.85 hp hr/gal.

Model 741 Gasoline (Test 779) with 251 cu in. displacement and 1700 rpm developed 57.79 hp and 9.91 hp hr/gal.

Model 731 Diesel (Test 743) with 4 1/8" x 5" bore and stroke 267 cu in. displacement and 1700 rpm developed 56.43 hp and 13.30 hp hr/gal.

J.I. Case 630 Diesel　　　788
1960-1964
48.24-40.67 hp @ 2000 rpm, 3 13/16" x 4 1/8";
188.4 cu in. 4 cyl Case engine;
12 forward gears, 1.5 to 20.2 mph;
12.85 hp hr/gal; 4637 lbs.

Model 630-C Diesel (Test 787) with same displacement and 2250 rpm developed 48.85 hp and 11.19 hp hr/gal.

J.I. Case 831-CK Diesel　　　917
1960-1969
64.26-58.68 hp @ 1900 rpm, 4 3/8" x 5";
301 cu in. 4 cyl Case engine;
8 forward gears, 1.7 to 17.6 mph;
12.23 hp hr/gal; 8070 lbs.

Model 841-CK Gasoline (Test 919) with 284 cu in. displacement and 1900 rpm developed 65.24 hp and 9.90 hp hr/gal.

Model 841-CK LPG (Test 921) with 284 cu in. displacement and 1900 rpm developed 64.08 hp and 7.48 hp hr/gal.

J.I. Case 841-C Gasoline 777
1960-1969

65.64-59.81 hp @ 1900 rpm, 4 1/4" x 5";
284 cu in. 4 cyl Case engine;
8 forward gears, 1.7 to 17.6 mph;
9.13 hp hr/gal; 7325 lbs.

Model 841-C LPG (Test 780) with same displacement and rpm developed 65.96 hp and 7.28 hp hr/gal.

Selective gear fixed ratio plus torque converter with lockout.

John Deere 2010 RU Gasoline 800
1960-1965

46.86-40.45 hp @ 2500 rpm, 3 5/8" x 3 1/2";
144.5 cu in. 4 cyl John Deere engine;
8 forward gears, 2.7 to 19.3 mph;
9.34 hp hr/gal; 5054 lbs.

Model 2010 Diesel (Test 799) with 165.1 cu in. displacement and 2500 rpm developed 46.67 hp and 10.51 hp hr/gal.

Selective gear with Partial Range Synchro-mesh.

John Deere 1010-RU Gasoline 802
1960-1965

36.13-30.62 hp @ 2500 rpm, 3 1/2" x 3";
115.5 cu in. 4 cyl John Deere engine;
5 forward gears, 2.6 to 16.9 mph;
7.99 hp hr/gal; 3923 lbs.

John Deere 1010-RU Diesel (Test 803) used a larger (3 5/8" x 3 1/2") bore and stroke, 144.5 cu in. displacement and 2500 rpm developing 35.99 hp and 10.66 hp hr/gal.

John Deere 3010 Gasoline 763
1960-1963

55.09-51.73 hp @ 2200 rpm, 4" x 4";
201 cu in. 4 cyl John Deere engine;
8 forward gears, 1.75 to 14.5 mph @ 1900 rpm;
9.12 hp hr/gal; 6320 lbs.

Model 3010 LPG (Test 764) with same displacement and rpm developed 55.39 hp and 7.34 hp hr/gal.

Model 3010 Diesel (Test 762) with 4 1/8" x 4 3/4" bore and stroke, and 254 cu in. displacement @ 2200 rpm developed 59.44 hp and 12.96 hp hr/gal.

Partial Range Synchro-mesh.

FARM TRACTORS 1950-1975

John Deere 4010 Diesel   761
1960-1963
84.00-73.65 hp @ 2200 rpm, 4 1/8" x 4 3/4";
380 cu in. 6 cyl John Deere engine;
8 forward gears, 1.5 to 14.25 mph; @ 1900 rpm;
12.38 hp hr/gal; 7445 lbs.

Model 4010 Gasoline (Test 759) with a 4" x 4" bore and stroke, 302 cu in. displacement and 2200 rpm developed 80.96 hp and 9.26 hp hr/gal.

Model 4010 LPG (Test 760) with 302 cu in. displacement and 2200 rpm developed 80.60 hp and 7.59 hp hr/gal.

Holder A-20 Diesel   No Test
1960-1967
24 engine hp @ 3000 rpm, 3.15" x 3.94";
61.31 cu in. 2 cyl MWM engine;
8 forward gears; 2569 lbs.

Four wheel drive, hydraulic steering.

Land Rover 88 Gasoline   749
1960-
30.90-28.09 hp @ 2000 rpm, 3.56" x 3.50";
139.5 cu in. 4 cyl Rover engine;
8 forward gears, 4 to 30 mph;
9.22 hp hr/gal; 3449 lbs.

Manufactured by the Rover Company, Solihull, Warwickshire, England.

Massey-Ferguson 65 Diesel   745
1960-1962
48.59-42.96 hp @ 2000 rpm, 3.6" x 5";
203.5 cu in. 4 cyl Perkins engine;
6 forward gears, 1.3 to 14.2 mph;
13.42 hp hr/gal; 4485 lbs.

1950-1975 FARM TRACTORS

Massey-Ferguson 88 Diesel 765
1960-1962
63.31-55.54 hp @ 2000 rpm, 4" x 5.5";
276.5 cu in. 4 cyl Continental engine;
8 forward gears, 1.7 to 21.5 mph;
12.00 hp hr/gal; 7165 lbs.

Essentially the same as Model 85.

McCormick-Farmall 504 Diesel 816
1960-1968
45.99-40.39 hp @ 2200 rpm, 3 11/16" x 4 25/64";
187.5 cu in. 4 cyl International Harvester engine;
10 forward gears, 1.3 to 17.6 mph;
12.30 hp hr/gal; 5605 lbs.

The first Hydrostatic power-steering agricultural tractor for International.

Model 504 Gasoline (Test 819) with 152.1 cu in. displacement (3.3/8" x 4 1/4" bore and stroke) and same rpm developed 46.20 hp and 9.89 hp hr/gal.

Model 504 LPG (Test 820) with same displacement and rpm as the 504 Gasoline developed 44.36 hp and 7.95 hp hr/gal.

These were the first International agricultural tractors to use dry type air cleaners.

McCormick-Farmall 404 Gasoline 818
1960-1967
36.70-33.47 hp @ 2000 rpm, 3 1/4" x 4 1/16";
134.8 cu in. 4 cyl International Harvester engine;
8 forward gears, 2.2 to 21 mph;
10.10 hp hr/gal; 4115 lbs.

First International 3-point hitch with draft control.

Minneapolis-Moline M-5 Diesel 758
1960-1963
58.15-50.99 hp @ 1500 rpm, 4 5/8" x 5";
336 cu in. 4 cyl Minneapolis-Moline engine;
10 forward gears, 1.7 to 17.4 mph;
12.94 hp hr/gal; 6965 lbs.

Available as Model M-602 (1963 to 1964).

Replaced the Model 5 Star.

FARM TRACTORS 1950-1975

Oliver 500 Gasoline   No Test
1960-1963
Manufacturer's estimate of 33-30 hp @ 2000 rpm, 3 1/2" x 4";
154 cu in. 4 cyl David Brown engine;
6 forward gears, 2.2 to 14.7 mph; 3750 lbs.

Available as diesel.

Oliver 1900 Diesel   768
1960-1964
89.35-80.42 hp @ 2000 rpm, 3 7/8" x 4 1/2";
212.4 cu in. (2 cycle) 4 cyl GM engine with blower;
6 forward gears, 1.5 to 13.4 mph;
12.52 hp hr/gal; 11,925 lbs.

The 1900 Series B introduced the Hydra-Power drive with partial range operator controlled power shifting.

Oliver 1800 Diesel   767
1960-1964
70.15-61.80 hp @ 2000 rpm, 3 7/8" x 4";
283 cu in. 6 cyl Oliver engine;
6 forward gears, 1.6 to 14.3 mph;
12.29 hp hr/gal; 8585 lbs.

Model 1800 Gasoline (Test 766) with 265 cu in. displacement (3 3/4" x 4" bore and stroke) and same rpm developed 73.92 hp and 10.99 hp hr/gal.

1950-1975 FARM TRACTORS

1961

The Turn to Turbochargers

By the 1960's, scientists and engineers had been experimenting with turbochargers for years in an effort to boost engine power. An original patent on turbocharging was granted in 1910 and in another decade it was being used on some aircraft engines. Turbochargers were used extensively on military aircraft during WW II. By the 1950's diesel truck engines with turbochargers were being used.

Turbochargers, driven by exhaust gases, compress incoming air above atmospheric pressure and force the air into combustion chambers. The boost in pressure may be in the range of 10 to 20 pounds per square inch at the intake manifold. This additional pressure, which may vary among different designs, makes the air more dense, allowing more fuel to be burned and thus develop more power from the engine and reduce exhaust noise.

It was no wonder that the Allis-Chalmers turbocharged D-19 tractor caused considerable excitement when it was introduced in 1961. It was the first diesel-powered farm tractor equipped with a turbocharger tested at Nebraska (Test 811).

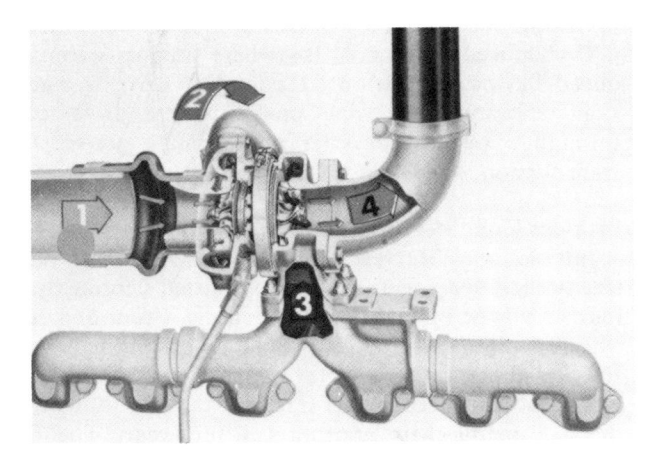

Dry Air Cleaner Used

The Allis-Chalmers Model D-19, in the 65-hp class, also was available for use with gasoline and LPG fuel. The D-19's were the first tractors tested at Nebraska equipped with dry-type air cleaners which used precleaners and automatic dust unloaders. However, they were not furnished with safety (or dual) elements at this time. Safety elements were developed later.

The new hydraulic transmissions continued to gain popularity as their potential advantages became apparent through practical use. They allowed automatic control of power and permitted engines to operate at their peak conditions despite variations in vehicle demands.

Try Gas Turbine

In its newest research tractor, the HT-340, International Harvester paired a gas turbine engine with hydrostatic transmission. The turbine engine was similar to the

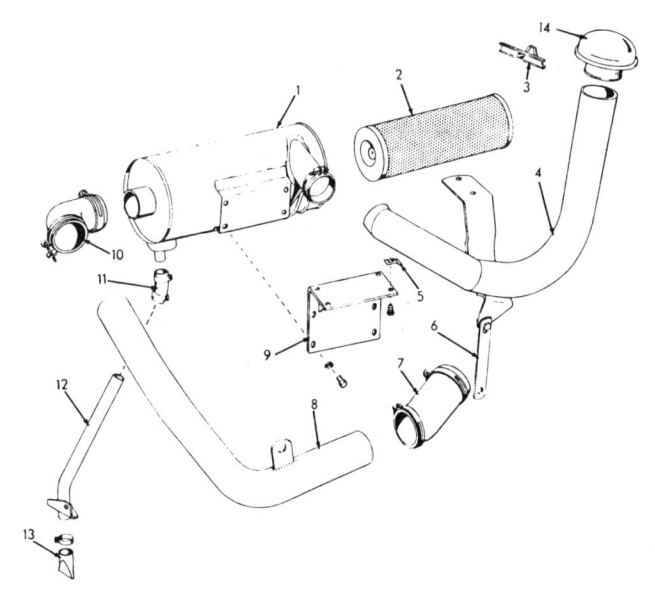

Turbocharger used on Allis-Chalmers D-19 tractor, believed to be one of the first diesel-powered farm tractors equipped with a turbocharger. Cutaway diagram shows (1) filtered air entering turbocharger through large tube, (2) turbocharger pumping air to intake manifold under high pressure, (3) exhaust gases driving turbocharger, and (4) turbocharger quieting exhaust noise without muffler.

Dry-type air cleaner used on the Allis-Chalmers D-19 tractors, with numbers indicating component parts.
1 Air cleaner body
2 Air cleaner element
3 Element lock
4 Air intake pipe
5 Screw support
6 Brace
7 Hose connection
8 Air pipe to engine
9 Air cleaner support
10 Rubber hose connection and clamps
11 Rubber hose connection
12 Pipe to automatic dust unloader
13 Automatic dust unloader
14 Air intake cap

1950-1975 FARM TRACTORS

In its research tractor, the HT-340, International Harvester paired gas turbine engine with hydrostatic transmission. Turbine engine permitted radically new visual concept. Skin sections are molded fiberglass.

turbo-prop engine used in some aircraft, but much smaller. The engine weighed only 60 lbs. It had a speed of 57,000 rpm and required much reduction gearing to drive the hydraulic pump. The pump, in turn, drove a hydraulic motor which propelled the tractor. The first public showing of this unit was at the Nebraska Power and Safety Day in Lincoln in 1961. The tractor now is on display at the Smithsonian Institution in Washington, DC.

Tractor a Real Giant

This same year International Harvester introduced its 4300 4-wheel drive tractor, reported to be the largest and most powerful wheel-type agricultural tractor on the market at the time. It was powered by a 300-hp, International 817 cu in. 6-cylinder, turbocharged diesel engine

Turbine engine used in International Harvester research tractor, the HT-340. Engine weighed only 60 pounds, with rpm of 57,000.

Above: Two turbochargers produced in 1954 and 1980 by The Garrett Corp., Los Angeles, California. Below: Turbocharger Technology Progress statistics represented by 16 years of turbocharger development. (Courtesy Garrett Airesearch)

-1954- PROTOTYPE OF FIRST PRODUCTION MODEL T15 TURBOCHARGER	-1980- PRESENT EQUIVALENT (AERODYNAMIC) PRODUCTION MODEL T04B TURBOCHARGER
AIRFLOW ..... 30-40 LBS/MIN PRESSURE RATIO .... 1.8 SPEED .... 40,000 RPM WEIGHT .... 94.5 LBS NUMBER OF PARTS 182	30-40 LBS/MIN ..... AIRFLOW 1.8 .... PRESSURE RATIO 85,000 RPM .... SPEED 17 LBS .... WEIGHT 53 .... NUMBER OF PARTS

operating at a governed speed of 2100 rpm. The 4300 was the first wheel-type agricultural tractor produced by International Harvester to use a turbocharger.

To utilize the power of this giant tractor, Harvester designed a 10-bottom fully-mounted plow that attached to the 3-point hitch. Field reports stated this tractor had enough power to pull this plow with ease. Total weight of the tractor, without ballast, was nearly 30,000 lbs, with the static weight split 63 percent on the front axle and 37 percent on the rear axle.

Other features of the 4300 included dry-type air cleaner, engine coolant heat exchanger for the crankcase oil, direct engine-driven hdyraulic pump, 8-speed transmission, and a 17-in. engine clutch operated by a foot pedal with air power assist. Air power also was used to operate the expanding brake shoes in all four wheels.

It should also be noted that these tractors were produced having an Allison "Torqmatic" drive transmission, transmitting engine power to the drive axles through a torque converter and 6-range power shift transmission. A cab was available if desired.

Provides Crab Steering

International Harvester's 4300 Diesel 4-wheel drive tractor included a feature which permitted steering by all four wheels or by just the front wheels. Front and rear wheels could be steered independently, providing a crab steering effect.

Two other International tractors, the 606 utility and the 414 made their appearances this year. The 606 gasoline and diesel models were in the 54-hp range. The 414 was in the 36-hp range with either gasoline or diesel models available.

The Case Churubusco Plant was closed in December, 1961. All production of Case crawler tractors had been consolidated into the Burlington, Iowa Plant. Rock Island production of small agricultural tractors had been

FARM TRACTORS 1950-1975

transferred to the Racine Clausen Tractor Works.

The year was also one in which the big tractors began to vie for the limelight. Ford came out with its largest and most powerful tractor to date—the Ford 6000. This tractor was powered by a 6-cylinder engine with 66 PTO horsepower. It was equipped with an upholstered seat and a tilt adjustable steering wheel for more operator comfort.

Allis-Chalmers D-19 Diesel 811
1961-1964
66.92-61.27 hp @ 2000 rpm, 3 9/16" x 4 3/8";
262 cu in. 6 cyl Allis-Chalmers engine;
8 forward gears, 1.9 to 13.9 mph;
10.76 hp hr/gal; 6835 lbs.

First agricultural tractor tested at Nebraska using a turbocharger to increase power.

Model D-19 Gasoline (Test 810) with same displacement and rpm developed 71.54 hp and 9.02 hp hr/gal.

Model D-19 LPG (Test 814) with same displacement and rpm developed 66.19 hp and 7.69 hp hr/gal.

David Brown 880 Diesel 902
1961-1964
40.42-35.77 hp @ 2200 rpm, 3 13/16" x 4 1/2";
154 cu in. 3 cyl David Brown engine;
6 forward gears, 2.3 to 15.4 mph;
13.33 hp hr/gal; 4315 lbs.

David Brown 990 Diesel 903
1961-1965
51.60-44.58 hp @ 2200 rpm, 3 5/8" x 4 1/2";
185.8 cu in. 4 cyl David Brown engine;
6 forward gears, 2.1 to 14.1 mph;
13.63 hp hr/gal; 4770 lbs.

Equipped with a differential lock in 1965.

1950-1975 FARM TRACTORS

Ford 6000 Diesel 783
1961-1967
66.17-61.31 hp @ 2400 rpm, 3.62" x 3.9";
241.7 cu in. 6 cyl Ford engine;
10 forward gears, 1.2 to 18.2 mph;
11.68 hp hr/gal; 7405 lbs.

Model 6000 Gasoline (Test 784) with 223 cu in. displacement and same rpm, (3.60 stroke) developed 66.86 hp and 7.66 hp hr/gal.

Heat exchanger in lower radiator to cool transmission oil, upholstered seat, with back rest, and hydraulic lift combined with accumulater.

Operator controlled full range power-shifting.

International B-414 Diesel 827
1961-1966
35.99-32.64 hp @ 2000 rpm, 3 1/2" x 4";
154 cu in. 4 cyl International Harvester engine;
8 forward gears, 1.5 to 15.5 mph;
12.50 hp hr/gal; 4050 lbs.

Model B-414 Gasoline (Test 817) with 143.1 cu in. (3 3/8" x 4" bore and stroke) and 2000 rpm developed 36.46 hp and 9.24 hp hr/gal.

Built in the United Kingdom. Replaced Model 275.

International 606 Gasoline 825
1961-1967
53.80-46.45 hp @ 2000 rpm, 3 9/16" x 3 11/16";
221 cu in. 6 cyl International Harvester engine;
10 forward gears, 1.3 to 17.7 mph;
9.81 hp hr/gal; 5215 lbs.

Model 606 Diesel (Test 826) with 236 cu in. displacement and 2000 rpm (3 11/16" x 3 11/16" bore and stroke) developed 54.31 hp and 12.63 hp hr/gal.

International 4300 Diesel 815
1961
214.23 db hp @ 2100 rpm, 5 3/8" x 6";
817 cu in. 6 cyl turbocharged International Harvester engine;
8 forward gears, 3.5 to 22.7 mph;
12.99 hp hr/gal; 29,815 lbs.

First International agricultural tractor equipped with a turbo-charger.

Massey-Ferguson Super 90 Diesel 822
1961-1965
68.53-61.54 hp @ 2000 rpm, 4.5" x 4.75";
302.2 cu in. 4 cyl Perkins engine;
8 forward gears, 1.7 to 21.6 mph;
13.21 hp hr/gal; 7245 lbs.

Available in gasoline and LPG models with Continental engines.

FARM TRACTORS 1950-1975

1962

White Purchases Cockshutt

On February 1, White Motor Corp. acquired the Cockshutt Farm Machinery Co. Ltd. and later incorporated a Canadian company under the name of Cockshutt Farm Equipment of Canada Ltd. At the same time, White traded Cockshutt's sales outlets in the US to Oliver in exchange for Oliver's outlets in Canada with the exception of Quebec where Co-op Federee maintained its outlets. White had acquired the Oliver Corp. and in 1963 acquired Minneapolis-Moline. To improve efficiency, the manufacturing of certain product lines was assigned to different plants. Combines were designed and built in Brantford and tractors at Charles City. Tillage and other tools were built in other US plants. In 1969, the Company officially became the White Farm Equipment, a wholly owned subsidiary of the White Motor Corp.

White Expands in Farm Field

The Oliver 600 complemented the Oliver 500 introduced in 1960. The 600 model was especially made for Oliver by David Brown. The Oliver 1600 Series was built in two series. All 1600's were powered with a choice of 6-cylinder engines using gasoline, LPG or diesel fuel. The first series of gasoline engines had 3 1/2 in. bore and 4 in. stroke. The second series gasoline and LPG engines had 3 5/8 in. bore and 4 in. stroke. Both series of diesel engines had 3 3/4 in. bore and 4 in. stroke. All 1600 tractors had a governed speed of 1900 rpm. Front-wheel drive was also available in the 1600 series tractors.

The Oliver 1900 Series B Diesel was the first Oliver tractor tested at Nebraska (Test 824) equipped with "Hydra-Power" drive permitting operator controlled partial range, power shifting on the go. When the operator shifted into Hydra-Power the travel speed was reduced about 26 percent and drawbar pull increased over 35 percent.

New Case Foot Clutch

A foot clutch was offered for the first time on the Case 930 Series tractors. A hand clutch option was still available until this model's discontinuance in 1967. The fuel tank on this model was also moved to a position behind the operator, allowing larger capacity and keeping the fuel away from engine heat.

In 1962, the Minneapolis-Moline G705 replaced the Model G-VI which continued in production until 1965. The G704 was the same as the G-VI except it had front-wheel drive. Model G706 Series was the same as the F705 except it was equipped with front-wheel drive, making it a 4-wheel drive tractor.

In 1962 the John Deere Tractor Company introduced the 5010 Model tractor. This was the first John Deere 2-wheel drive tractor that had over 100 horsepower at the drawbar. The PTO shaft was intended only for 1000 rpm due to the high horsepower of this tractor. The 540 rpm PTO was not suitable for such power.

Allis-Chalmers D-10 Gasoline   812
1962-1967
33.46-29.97 @ 1650 rpm, 3 1/2" x 3 7/8";
149 cu in. 4 cyl Allis-Chalmers engine;
4 forward gears, 2.0 to 11.4 mph;
11.12 hp hr/gal; 3000 lbs.

1950-1975 FARM TRACTORS

Allis-Chalmers (Series II) D-12 Gasoline 813
1962-1964
33.32-29.90 hp @ 1650 rpm, 3 1/2" x 3 7/8";
149 cu in. 4 cyl Allis-Chalmers engine;
4 forward gears, 2.0 to 11.4 mph;
10.87 hp hr/gal; 3050 lbs.

Ford 2000 Super Dexta Diesel 844
1962-1964
38.83-32.89 hp @ 2250 rpm, 3.6" x 5";
152.7 cu in. 3 cyl Ford engine;
6 forward gears, 1.6 to 19.5 mph;
12.45 hp hr/gal; 3510 lbs.

Imported from the United Kingdom.

John Deere 5010 Diesel 828
1962-1965
121.12-108.91 hp @ 2200 rpm, 4 3/4" x 5";
531 cu in. 6 cyl John Deere engine;
8 forward gears, 1.75 to 15.5 mph @ 1900 rpm;
12.94 hp hr/gal; 12,925 lbs.

Partial Range Syncro-Mesh transmission.

Kramer KL 400 Diesel 821
1962
32.66-28.36 hp @ 2300 rpm, 3.74" x 4.72";
155.5 cu in. 3 cyl Deutz air-cooled engine;
10 forward gears, 0.9 to 12.1 mph;
11.40 hp hr/gal; 4225 lbs.

Kramer-Werke, Uberlingen, Bodensee, West Germany.

FARM TRACTORS 1950-1975

Minneapolis-Moline G705 Diesel 835
1962-1965
101.01-93.01 hp @ 1600 rpm, 4 5/8" x 5";
504 cu in. 6 cyl Minneapolis-Moline engine;
5 forward gears, 3.3 to 18.3 mph;
12.63 hp hr/gal; 8155 lbs.

Model G705 LPG (Test 836) was essentially the same as the G705 Diesel except 7.91 hp hr/gal.

Oliver 600 Diesel No Test
1962-1963
Manufacturer's estimate of 48-43 hp @ 2000 rpm;
David Brown engine.

Built especially for Oliver by David Brown, England.

Minneapolis-Moline G706 Diesel 833
1962-1965
101.68-89.13 hp @ 1600 rpm, 4 5/8" x 5";
504 cu in. 6 cyl Minneapolis-Moline engine;
5 forward gears, 3.3 to 18.3 mph;
12.05 hp hr/gal; 9165 lbs.

Model G706 LPG (Test 834) was essentially the same except for 7.82 hp hr/gal.

Oliver 1600 Diesel 840
1962-1964
57.95-48.78 hp @ 1900 rpm, 3 3/4" x 4";
265 cu in. 6 cyl Oliver engine;
12 forward gears, 1.9 to 13.6 mph;
10.76 hp hr/gal; 7480 lbs.

Model 1600 Gasoline (Test 841) with 231 cu in. displacement (3 1/2" x 4" bore and stroke) and same rpm developed 56.50 hp and 8.36 hp hr/gal.

Oliver 1800 Series B 4-wheel drive Gasoline   846
1962-1964
80.72-69.52 hp @ 2200 rpm, 3 7/8" x 4";
283 cu in. 6 cyl Oliver engine;
12 forward gears, 1.2 to 14.4 mph;
9.83 hp hr/gal; 10,155 lbs.

Model 1800 Series B 4-wheel drive Diesel (Test 832) with 310 cu in. displacement (3 7/8" x 4 3/8" bore and stroke) and same rpm developed 76.97 hp and 10.94 hp hr/gal.

1963

Into the 4-Wheel Drives

Horsepower continued to go up and up with each succeeding year as tractor manufacturers announced their new models. And as horsepower increased, so did the pull required by each rear wheel.

To compensate for this situation, suppliers began to provide auxiliary front wheels that could be driven from the PTO or by hydraulic power. This, in practice, made a 2-wheel drive tractor into a 4-wheel drive tractor.

The introduction of 4-wheel drive tractors, with independent power to each wheel, brought on a variety of steering combinations. One company which recognized the possibilities offered with this situation was the M-R-S Manufacturing Company, Flora, Mississippi.

This company began by building industrial tractors and heavy earth-moving equipment in Jackson, Mississippi in 1943. As the demand for horsepower increased, M-R-S developed a 4-wheel drive tractor featuring independently controlled "All-Wheel Steering." This made possible independent front-wheel steering, independent rear-wheel steering, coordinated steering, and crab steering.

M-R-S introduced its 4-wheel drive farm tractor, Model A-80, in 1963 with a 141-hp rating, and in subsequent years it was to produce additional models with greater horsepower.

Power Train Redesigned

International Harvester chose this year to make major changes in its tractor design with the introduction of its 706 and 806 Series. The new line had a completely redesigned power train and hydraulic system. Other major changes included the following:

Operator platform cleared of obstructions by moving shift levers to side of steering wheel support column.

Seat moved ahead of rear axle to provide a smoother ride.

New 4-speed sliding gear transmission followed by a 2-speed range section, providing 16 forward and eight reverse speeds when equipped with TA. Pressure lubrication to transmission and final drive components.

New hydraulic draft control, 3-point hitch with torsion bar lower link sensing.

Hydrostatic steering.

Hydraulically actuated self-adjusting brakes.

Cooler for transmission and hydraulic oil.

Power shift, dual speed (540 and 1000 rpm) IPTO (Independent PTO).

Dry-type air cleaners.

Dual rear wheels.

A factory-installed front-wheel drive attachment for 706 and 806 tractors.

Into Big Tractor Market

Allis-Chalmers' entry into the big tractor market was its D-21, with an excess of 100 horsepower. This tractor had a 52-gal fuel tank, a 4-position steering wheel, redesigned instrument panel, bigger fenders, an under-the-hood aluminized muffler, an alternator instead of a generator, and two 12-volt batteries for easier starting. This was the first Allis-Chalmers tractor equipped with a 1000 rpm PTO and the first to be equipped with hydrostatic power steering.

John Deere added a power shift transmission to its new 3020 and 4020 models. This transmission made it possible for the operator to shift through all eight speeds with one hand lever, under full power. This was the last year John Deere built tractors to use LPG, an indication that LPG was losing out to diesel as a fuel for agricultural tractors.

The 4020 model offered power front-wheel drive to provide better traction under adverse conditions. The 4020 also had stronger and longer rear axles to make a better and safer mounting for dual wheels.

Massey-Ferguson/Minneapolis-Moline Connection

In 1958 Massey-Ferguson obtained a large tractor from Minneapolis-Moline. This tractor was sold as the Massey-Ferguson 95 in Canada only (some of these tractors may have come back over the border). Later the model designation was changed to Massey-Ferguson 95 Super. Except for sheet metal, paint and decals, these tractors were basically the Minneapolis-Moline Models GBD and GVI.

The Model 97 tractor was basically a Minneapolis-Moline G705, G706, G707 or G708 tractor. The G705 and G707 were 2-wheel drive tractors. G707 was an updated version of the G705. G706 and G708 were equipped with power front wheels. G708 was an updated version of the G706 tractor. The Massey-Ferguson 97 tractor was available as a 2-wheel drive tractor or a factory installed power front-wheel drive.

All Perkins engines now were being equipped with direct injection combustion chambers. Previously all Perkins engines used precombustion chambers or energy cells. All French built Massey-Ferguson 25 tractors were being equipped with diesel engines.

Second Generation Started

The Steiger Brothers started the second generation of new models with the 1200, 1700, 2200, and 3300. The 1200 model used a 4-cylinder 212 cu in. Detroit Diesel engine. It had 12 forward gears, compared to five for its predecessor, the Steiger No. 1 model.

The Model 3300 was the largest and most powerful of the Steiger tractors so far, with a 318-hp V-8 Detroit Diesel engine. The engine ran at 2300 rpm and the tractor had 16 forward speeds. About 120 of these new models were built and sold during the 1960's, mostly in Canada, Montana, and the Dakotas.

New Injection System for Diesels

It was now becoming evident that diesel engines would one day power a great many of America's farm tractors. But the fact was not so evident back in 1947 when diesel power still had limited use. That was the year in which Vernon Roosa, who had been installing and maintaining diesel-electric generator sets in New York City, discarded the traditional in-line injection pump with a pumping element for each cylinder and substituted a single pumping unit to feed all the cylinders.

The Roosa Master type diesel fuel injection pump simplified the pumping units and reduced their size and cost. Undoubtedly, the development of Roosa's fuel injection system accelerated the use of diesel-powered tractors in both industry and agriculture.

This photograph compares Roosa Master Pencil Nozzles with the conventional 21 mm nozzle. Note that the pencil nozzles were smaller. They also were more versatile and well suited to the smaller direct-injection diesels of the 1960's. The variations in the pencil nozzles shown indicate their development for different diesel engines. (Courtesy Stanadyne Diesel Systems)

In September, 1963, Roosa came out with still another development in fuel injection systems. It was the Roosa Master Pencil Nozzle. This nozzle was smaller, less expensive, and more versatile than the traditional nozzle and was better suited to the smaller direct-injection diesels of the 1960's.

Tricycle Type Tractor Popularity Fades

Through the years, two basic tractor models—the row-crop and the standard or wheatland—had dominated the agricultural scene. Now the row-crop, often referred to as the tricycle type, was losing out to the standard versions. Users were finding the wide front end, with space adjustments, more to their liking. The wide front ends of the standard models permitted larger wheels in front with larger tires to carry heavier loads. Some operators also thought tractors with wide front ends were safer—not so likely to tip over.

One additional historical note for the year 1963: the White Motor Corp. acquired controlling interest in the Minneapolis-Moline Company, Hopkins, Minnesota.

The in-line type of diesel fuel injection system, using a separate pump for each cylinder, had been generally standard for many years. Instead of a separate pumping unit for each cylinder the Roosa Master distributor type, developed in 1947, used a single pumping unit. The Roosa Master type of diesel fuel injection pump simplified the construction of pumping units, reduced the size and materially reduced the cost. (Courtesy Stanadyne Diesel Systems)

Allis-Chalmers D-21 Diesel   855
1963-1969

103.06-94.05 hp @ 2200 rpm, 4 1/4" x 5";
426 cu in. 6 cyl Allis-Chalmers engine;
8 forward gears, 1.6 to 16.2 mph;
13.26 hp hr/gal; 10,745 lbs.

John Deere 3020 Gasoline   851
1963-1972

64.14-55.43 hp @ 2500 rpm, 4 1/4" x 4";
227 cu in. 4 cyl John Deere engine;
8 forward gears, 1.5 to 16.5 mph @ 2100 rpm;
7.85 hp hr/gal; 7695 lbs.

Model 3020 LPG (Test 852) with same displacement and rpm as gasoline model developed 64.70 hp and 6.76 hp hr/gal.

Model 3020 Diesel (Test 848) with 4.75 stroke and 270 cu in. displacement and same rpm developed 65.28 hp and 10.61 hp hr/gal.

Transmission fixed ratio operator controlled full range power-shifting.

John Deere 4020 Diesel   849
1963-1972

91.17-78.04 hp @ 2200 rpm, 4.25" x 4.76";
404 cu in. 6 cyl John Deere engine;
8 forward gears, 1.5 to 15.5 mph @ 1900 rpm;
11.77 hp hr/gal; 8945 lbs.

Model 4020 Gasoline (Test 850) with 340 cu in. displacement (4 1/4" x 4" bore and stroke) and same rpm developed 88.09 hp and 7.79 hp hr/gal.

Model 4020 LPG (Test 853) with 340 cu in. displacement and same rpm developed 90.48 hp and 7.42 hp hr/gal.

Operator controlled full range power-shifting.

John Deere 4020 Power Shift Diesel   1024
1969-

95.83-84.13 hp @ 2200 rpm, 4.25" x 4.75";
404 cu in. 6 cyl John Deere engine;
8 forward gears, 1.8 to 18.7 mph;
12.50 hp hr/gal; 9560 lbs.

1950-1975 FARM TRACTORS

International Farmall 706 Diesel   856
1963-1967
72.42-67.55 hp @ 2300 rpm, 3 11/16" x 4 25/64";
282 cu in. 6 cyl International Harvester engine;
16 forward gears, 1.2 to 18.4 mph;
12.01 hp hr/gal; 8530 lbs.

Model 706 Gasoline (Test 858) with 263 cu in. displacement (3 9/16" x 4 25/64" bore and stroke) and same rpm developed 73.82 hp and 10.45 hp hr/gal.

Model 706 LPG (Test 860) 263 cu in. displacement and same rpm developed 73.66 hp and 8.02 hp hr/gal.

Redesigned power train and hydraulic system.

Massey-Ferguson 97 Diesel   835
(Also M-M G 705 Diesel)
1963-1965
101.01-93.01 hp @ 1600 rpm, 4 5/8" x 5";
504 cu in. 6 cyl Minneapolis-Moline engine;
5 forward gears, 3.3 to 18.3 mph;
12.63 hp hr/gal; 8155 lbs.

Model 97 LPG (Test 836) with same displacement and rpm developed 101.05 hp and 7.91 hp hr/gal (also M-M G 705 LPG).

Massey-Ferguson 25 Diesel   854
1963
24.39-20.58 hp @ 2000 rpm, 3 1/8" x 3 1/2";
107 cu in. 4 cyl Perkins engine;
8 forward gears, 1.1 to 14.1 mph;
11.78 hp hr/gal; 3115 lbs.

Built in France.

M-R-S A-80 Diesel   No Test
1963
141 hp @ 2100 rpm, 4 1/4" x 5";
284 cu in. 4 cyl 2 cycle GM engine;
10 forward gears, 1.8 to 16.3 mph;
17,000 lbs.

All wheel drive.

FARM TRACTORS 1950-1975

Oliver 1800 Series B Diesel 832
1962
76.97-64.04 hp @ 2200 rpm, 3 7/8" x 4 3/8";
310 cu in. 6 cyl Oliver engine;
12 forward gears, 1.1 to 14.4 mph;
10.94 hp hr/gal; 10,675 lbs.

The 1800 Series B 4-wheel drive Gasoline (Test 846) with 283 cu in. displacement and same rpm developed 80.72 hp and 9.83 hp hr/gal.

Steiger 1700 Diesel No Test
1963-1969
Advertised @ 216 hp @ 2800 rpm, 3.875" x 4.5";
318 cu in. V6 cyl Detroit Diesel engine;
9 forward gears, 1.8 to 19 mph;
19,000 lbs.

Steiger 2200 Diesel No Test
1963-1969
Advertised @ 265 hp @ 2300 rpm, 4 1/4" x 5";
426 cu in. V6 cyl Detroit Diesel engine;
9 forward gears, 1.9 to 21 mph;
24,000 lbs.

Steiger 1200 Diesel No Test
1963-1969
3.875" x 4.5";
212 cu in. 4 cyl Detroit Diesel engine;
12 forward gears.

1950-1975 FARM TRACTORS

Steiger 3300 Diesel   **No Test**
1963-1969
Manufacturer's estimate of 318 hp @ 2300 rpm, 4 1/4'' x 5'';
568 cu in. V8 cyl Detroit Diesel engine;
16 forward gears, 2.7 to 22.7 mph;
31,000 lbs.

Operator-controlled full range power-shift transmission.

1964

Policy to Certify Horsepower

The Oliver 1650 tractor was the first of the 1250, 1550, 1650, 1750, 1850 and 1950 Series, with its introduction, "Certified Horsepower" was born. This new method of quoting horsepower capabilities involved testing each production tractor and certifying that horsepower met minimum requirements. A decal with horsepower written on it is affixed to each tractor. The new 1650 Model tractor was powered by a choice of 6-cylinder engines using gasoline, LPG or diesel operating at 2200 rpm.

The new features of the 1650 Series were independent dual PTO (540 and 1000 rpm), super deluxe seat, "Dry type" air cleaner, speed-o-tac, tilt-telescope power steering, hydra-lectric hydraulics, 3-point hitch and 4-wheel drive option.

The 1850 was second in the 1250, 1550, 1650, 1850 and 1950 Series. All 1850's were powered by 6-cylinder gasoline, LPG or diesel engines operating at 2400 rpm. The 1850 was the first Oliver tractor tested at Nebraska using hydrostatic power steering. This tractor also had many of the same features as found in the 1650 model.

When the 1850 Model tractor was introduced in 1964, Oliver updated many 1800 "C" tractors to obtain a higher performance. These updated tractors were known as the Oliver 1750.

The Oliver Corporation introduced "certified horsepower" in 1964. A certified sticker was attached to every tractor to give the buyer confidence in its performance.

Ford Expands Market

Ford continued to expand its worldwide market by opening manufacturing and assembly operations at plants in New Basildon, England and Antwerp, Belgium. The Highland Park plant in Detroit was renovated to assemble components for its tractors intended for the American market.

Minneapolis-Moline replaced the 602 Series with the 670 Series. Both tractor engines had the same displacement and rpm. The main difference was the appearance and improved hydraulics. These tractors were all available to use gasoline, LPG and diesel fuels.

In 1964 the U-302 Series was introduced replacing the 4 Star Series. The Model U-302 used diesel or LPG engines. This new model was the first Minneapolis-Moline tractor engine equipped with a dry type air cleaner.

Allis-Chalmers added console-mounted controls, a 12-position operator seat, and an adjustable steering wheel to its new One-Ninety 6-cylinder diesel. Other features included a power clutch and either 540 or 1000 rpm for the PTO. The tractor had a roll-shift front axle and power-shift rear wheels.

First 4-Wheel Drive Case

Case introduced their first 4-wheel drive tractor— the Model 1200 Traction King Diesel. This was the first Case agricultural tractor equipped with a 451 cu in. displacement engine with a turbocharger. With increased fuel consumption and more power there was more heat to dissipate, therefore, this engine was equipped with a crankcase oil-to-engine coolant heat exchanger.

The new Case 1200 Diesel was equipped with vacuum-assisted hydraulic brakes in each drive wheel. Hydrostatic power steering was used for the two front wheels through the normal steering wheel and a separate hydraulic power steering was used to control the rear drive wheels via a hand lever.

The Eagle Hitch was discontinued with the 730 and 830, but was re-introduced with a new 3-point hitch and hydraulic system with draft control called Draft-O-Matic, and several other refinements. The Draft-O-Matic system utilized "wishbone" leaf springs for lower link load sensing and a complex mechanical linkage to transmit the spring movement to a valve. A hand lever provided multiple draft control settings from none (depth control) to the maximum sensitivity available. A

1950-1975 FARM TRACTORS

second hand lever provided raise, lower, float, and depth-positioning functions.

A 930 General Purpose model was also introduced this year. This model featured a larger 8-speed transmission, dry clutch, choice of gasoline, diesel, or LPG engines, rear mounted fuel tank, and the new Draft-O-Matic 3-point hitch system. This tractor was very different from the 930 model with 6-speed transmission, wet clutch, and roller chain final drive which was still produced until 1967.

Case dedicated a new modern and efficient research and test center in Racine. It was also in 1964 that Case acquired the Colt Manufacturing Co., Inc. of Winneconne, Wisconsin, as a wholly-owned subsidiary for small lawn and garden tractors in the 10-12 hp class.

While Case was acquiring Colt, the Kern County Land Company was acquiring a controlling interest in Case. This resulted in a refinancing plan that built a sound basis for the future operation of the company.

F W D Wagner WA4 Diesel   864
1964
97.30 db hp @ 2500 rpm, 3 7/8'' x 4 1/2'';
212.3 cu in. 2 cycle 4 cyl GM engine;
8 forward gears, 1.6 to 18.3 mph;
12.00 hp hr/gal; 11,035 lbs.

J.I. Case 1200 Diesel   868
1964-1968
119.90-106.86 hp @ 2000 rpm, 4 3/8'' x 5'';
451 cu in. 6 cyl turbocharged Case engine;
6 forward gears, 2.5 to 14.1 mph;
11.81 hp hr/gal; 16,585 lbs.

First four-wheel drive Case tractor tested at Nebraska.

This tractor was equipped with hydraulic brakes in each wheel. Front and rear wheels could be steered independently.

First Case agricultural tractor equipped with a turbocharger.

International 424 Diesel   911
1964-1968
36.91-32.57 hp @ 2000 rpm, 3 1/2'' x 4'';
154 cu in. 4 cyl International Harvester engine;
8 forward gears, 1.5 to 15.5 mph;
11.57 hp hr/gal; 4055 lbs.

Model 424 Gasoline (Test 908) with 145.3 cu in. displacement (3 3/8'' x 4 1/16'' bore and stroke) developed 36.97 hp and 9.57 hp hr/gal.

Replaced Model B 414.

FARM TRACTORS 1950-1975

Minneapolis-Moline U302 Gasoline   862
1964-1970
55.82-48.91 hp @ 1900 rpm, 3 3/4" x 5";
221.1 cu in. 4 cyl Minneapolis-Moline engine;
10 forward gears, 1.8 to 19.5 mph;
9.56 hp hr/gal; 5475 lbs.

Model U302 LPG (Test 863) with same displacement and rpm developed 55.69 and 8.03 hp hr/gal.

First Minneapolis-Moline tractors tested at Nebraska with dry type air cleaning.

Minneapolis-Moline M670 LPG   924
1964-1970
74.16-64.70 hp @ 1600 rpm, 4 5/8" x 5";
336 cu in. 4 cyl Minneapolis-Moline engine;
10 forward gears, 1.6 to 17.2 mph;
8.27 hp hr/gal; 7545 lbs.

Model M670 Gasoline (Test 925) with same displacement and rpm developed 73.02 hp and 9.23 hp hr/gal.

Also available as Model M670 Super.

Model M670 replaced Model 602 series. Main difference was the appearance and hydraulic equipment.

Nuffield 10/60 Diesel   907
1964-1967
55.36-47.94 hp @ 2000 rpm, 3.94" x 4.72";
230 cu in. 4 cyl BMC engine;
10 forward gears, 1.4 to 17.4 mph;
12.81 hp hr/gal; 5865 lbs.

Three earlier models were tested:

1950-Nuffield Universal Three Diesel (Test 663) 34.31 max hp and 13.19 hp hr/gal.

1962-Nuffield 460 Diesel (Test 809) 54.90 max hp and 13.85 hp hr/gal.

1965-Nuffield 10/42 Diesel (Test 905) 39.06 max hp and 13.06 hp hr/gal.

Oliver 1650 Diesel   873
1964-1969
66.28-57.16 hp @ 2200 rpm, 3 7/8" x 4";
283 cu in. 6 cyl Oliver engine;
12 forward gears, 1.9 to 13.7 mph;
10.45 hp hr/gal; 7815 lbs.

Model 1650 Gasoline (Test 874) with 265 cu in. displacement (3 3/4" x 4" bore and stroke) and same rpm developed 66.72 hp and 9.32 hp hr/gal.

1950-1975 FARM TRACTORS

Oliver 1950 Diesel 871
1964-1970
105.79-99.00 hp @ 2400 rpm, 3 7/8" x 4 1/2";
212.4 cu in. 2 cycle 4 cyl GM engine with blower;
12 forward gears, 1.3 to 15.6 mph;
13.03 hp hr/gal; 11,955 lbs.

First Oliver tested with dry (dual type) air cleaner.

Model 1950 4-wheel drive Diesel (Test 872) with same displacement and rpm, slightly less drawbar power and 12.51 hp hr/gal.

Zetor 4011 Diesel 867
1964
48.50-41.84 hp @ 2000 rpm, 3.74" x 4.33",
190.5 cu in. 4 cyl Zetor engine;
0.7 to 15.9 mph;
14.03 hp hr/gal; 4910 lbs.

Four other Zetor tractors were tested:

Zetor 50 Super Diesel (Test 748) 49.47 hp and 14.29 hp hr/gal.

Zetor 3011 Diesel (Test 823) 33.66 hp and 13.06 hp hr/gal.

Zetor 2011 Diesel (Test 866) 21.20 hp and 12.12 hp hr/gal.

Zetor 5511 Zetormatic Diesel (Test 960) 50.87 hp and 13.42 hp hr/gal.

All manufactured at Brno, Czechoslovakia

1965

Better Hitches Developed

By the mid-1960's nearly all the major agricultural tractor manufacturers had made or were making refinements in hitches. Many of the new hitch arrangements were designed to integrate the tractor and implement to function as a single operating unit. In effect, the implement became self-propelled with the advantages of ease of control, plus a reduction in operating costs claimed by the manufacturer.

Massey-Ferguson heralded its new Pressure Control hitches as ushering in a new era of hydraulic weight transfer for pull-type implements, as well as 3-point hitch implements having gauge wheels. The firm made available two types of pressure control hitches—the cone and socket, and the universal.

With the cone and socket hitch, the tractor was backed into a cone on the implement tongue and locked with a turn of the attaching collar. With the universal hitch, the implement tongue was attached to the tractor drawbar, then a chain was inserted through a loop on the implement tongue, pulled up snug and fastened to the chain claw.

A pressure control lever on the hydraulic control quadrant made it possible for the operator to transfer weight from the implement and tractor front end to the rear wheels.

The company claimed pressure control hitching reduced operating costs by allowing a lighter tractor to do the work of a heavier one, which would require more fuel, by utilizing and properly distributing the available weight of both the implement and the tractor.

Major Company Developements

The International Harvester Company presented three new farm tractor models—the 1206, 4100, and the 656 line. The 1206 was Harvester's first Farmall having more than 100 PTO horsepower. It was powered by a 6-cylinder, 361 cu in. turbocharged diesel engine.

The International 1206 and 4100 diesel tractors were the first International tractors to be equipped with dual-type dry air cleaners. The dual-type dry air cleaners were equipped with a safety element inside the main element.

The company also offered a factory-installed cab with heater and air conditioner. This marked the beginning of factory-installed cabs on 2-wheel drive International tractors. Stolper-Allen, Menomonee Falls, Wisconsin, manufactured the cab.

Minneapolis-Moline's answer to the 100-hp models of competitors was the G1000. It was available with either LPG or diesel engines and replaced the 706 and 705 models.

Allis-Chalmers followed its One-Ninety model with a One-Ninety XT. The new diesel tractor was similar to the One-Ninety, but equipped with a turbocharger that added 16 hp to engine output.

The Oliver 1550 tractor offered a choice of a 6-cylinder gasoline, LPG or diesel engine, operating at 2200 rpm. This tractor had many of the same features found on the earlier versions of this series.

Also abroad, David Brown produced improvements in its 770 diesel, with two shift levers providing 12 forward travel speeds in a new transmission.

The new Massey-Ferguson 135 replaced the Model 35 with more power as the 165 replaced the 65. The 135 was the first Massey tractor tested at Nebraska equipped with a dry type air cleaner.

About this time roll-over protection was becoming available for Massey-Ferguson tractors.

Along with the new Model 175 announced in 1965, a new type of draft control had been announced for trailing implements. The 150, the 180 and the 1100 were all new models introduced the same year. The 150 replaced the earlier Model 50. The new 1100 model was approaching the 100 hp size in the ever increasing horsepower race. The Category III hitch was now available in the Model 1100 tractor. The 1100 was the first Massey-Ferguson tractor to be equipped with a radiator located in front of the regular radiator to cool the transmission and hydraulic oil.

In an era when greater attention was being given to safety, John Deere developed their Roll-Gard rollover protection, more commonly known as "ROPS". The new John Deere 5020 Diesel replaced the earlier 5010 model with additional horsepower.

First Japanese Tractor Appears

The first Japanese-built agricultural tractors appeared in the US with the premier showing of the Kubota RV gasoline tractor, and the Kubota D 20 diesel, produced by the Kubota Iron and Machinery Works, Ltd., Osaka, Japan.

The Kubota RV had a single cylinder engine which developed nine horsepower, maximum belt. The Kubota D 20 produced 20 horsepower.

Another foreign-made small tractor made its appearance this year. It was the Oliver 1650, a 4-cylinder tractor manufactured by the Fiat Company of Italy and adapted to Oliver specifications. The gasoline engine for this tractor had a certified horsepower of 35 PTO and the diesel engine a certified horsepower of 38. Both the gasoline and diesel engines operated at 2500 rpm.

Work Toward Safety

While the race toward bigness appeared to dominate the farm tractor scene, many manufacturers, institutions, and research centers continued efforts to make big and little tractors alike safer vehicles for the operator.

Case started mounting fuel tanks behind the driver's seat. John Deere, Massey-Ferguson, and other firms were working on roll-over protection, with special guards and reinforced frames. And the Nebraska Tractor Testing Laboratory was testing the first tractor with a roll-over protective cab.

Increased movement of tractors and implements on public roads focused attention on collisions of automobiles and trucks with agricultural equipment. Many rear-end accidents resulted when motorists misjudged the slow speed at which farm vehicles ahead were traveling. Safety specialists, engineers, and law enforcement officials recognized this problem and began to consider solutions.

The Automotive Safety Foundation provided the Agricultural Engineering Department at The Ohio State University with a grant to develop a suitable warning emblem for slow-moving vehicles. Under the leadership of Kenneth Harkness, then a graduate student at Ohio State, there evolved a triangular-shaped fluorescent yellow and orange emblem which could be mounted on the rear of slow-moving vehicles traveling on the highway.

Ohio and Nebraska were among the first states to pass legislation requiring that slow-moving vehicles display the emblem. Later, most state legislatures were to pass similar laws, and the emblem was to be accepted throughout the country as a universal symbol for vehicles moving at slow speeds on public roads.

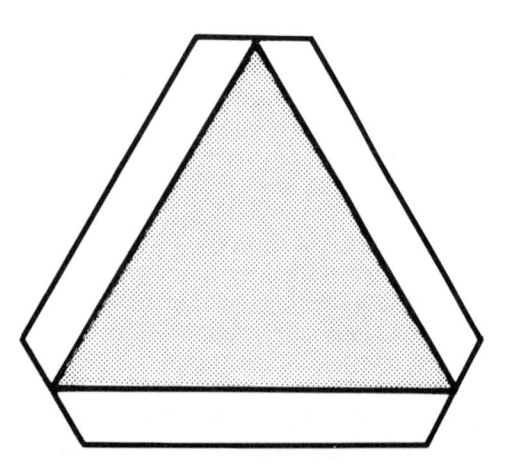

Allis-Chalmers (Series II) D-21 Diesel   904
1965-1969
127.75-117.57 hp @ 2200 rpm, 4 1/4" x 5";
426 cu in. 6 cyl turbocharged Allis-Chalmers engine;
8 forward gears, 1.6 to 16.2 mph;
13.63 hp hr/gal; 10,675 lbs.

Allis-Chalmers One-Ninety XT Gasoline   929
1965-1970
89.53-81.26 hp @ 2200 rpm, 3 7/8" x 4 1/4";
301 cu in. 6 cyl Allis-Chalmers engine;
8 forward gears, 2.1 to 13.6 mph;
8.73 hp hr/gal; 7675 lbs.

Model One-Ninety Gasoline (Test 928) was similar to the XT Model except smaller displacement of 265 cu in. (3 3/4" x 4" bore and stroke) and same rpm which developed 75.37 hp and 8.20 hp hr/gal.

Model One-Ninety XT LPG (Test 927) with same rpm and 301 cu in. displacement developed 85.25 hp and 7.30 hp hr/gal.

Slow Moving Vehicle (SMV) symbol.

FARM TRACTORS 1950-1975

Allis-Chalmers One-Ninety XT Diesel   887
1965-1972
93.64-84.06 hp @ 2200 rpm; 3 7/8" x 4 1/4";
301 cu in. 6 cyl turbocharged Allis-Chalmers engine;
8 forward gears, 2.1 to 13.6 mph;
13.07 hp hr/gal; 7945 lbs.

Model One-Ninety Diesel without turbocharger (Test 886) with same displacement and rpm developed 77.20 hp and 11.73 hp hr/gal.

David Brown Selectomatic 880 Diesel   946
1965-1971
42.29-36.37 hp @ 2200 rpm, 3.94" x 4.5";
164.4 cu in. 3 cyl David Brown engine;
6 forward gears, 2.3 to 15.4 mph;
12.70 hp hr/gal; 4470 lbs.

David Brown Selectamatic 770 Diesel   945
1965-1970
32.12-28.37 hp @ 2000 rpm, 3.94" x 4.0";
146.1 cu in. 3 cyl David Brown engine;
12 forward gears, 1 to 15.4 mph;
14.09 hp hr/gal; 3780 lbs.

David Brown Selectomatic 990 Diesel   977
1965-1971
52.07-44.83 hp @ 2200 rpm, 3.94" x 4.0";
194.9 cu in. 4 cyl David Brown engine;
6 forward gears, 2 to 13.8 mph;
13.98 hp hr/gal; 4940 lbs.

1950-1975 FARM TRACTORS

David Brown Selectomatic 3800 Gasoline   1018
1965-1971
39.16-34.14 hp @ 2200 rpm, 3.94" x 4";
146.1 cu in. 3 cyl David Brown engine;
6 forward gears, 2.5 to 16.7 mph;
8.77 hp hr/gal; 3885 lbs.

J.I. Case 941 GP LPG   922
1964-1969
85.86-75.61 hp @ 1800 rpm, 4" x 5";
377 cu in. 6 cyl Case engine;
8 forward gears, 1.8 to 14.7 mph;
7.04 hp hr/gal; 8975 lbs.

This model was the last LPG tractor built by J.I. Case.

Model 941 GP Gasoline (Test 920) with same displacement and rpm developed 86.21 max hp and 8.79 hp hr/gal.

Model 931 GP Diesel (Test 918) with 401 cu in. displacement and same rpm developed 85.39 max hp and 12.04 hp hr/gal.

John Deere 1020 Gasoline   935
1965-1972
38.82-32.74 hp @ 2500 rpm, 3.86" x 3.86";
135.3 cu in. 3 cyl John Deere engine;
8 forward gears, 1.4 to 16.6 mph;
8.08 hp hr/gal; 4730 lbs.

By increasing the stroke to 4.33, and the engine displacement to 152 cu in., the 1020 Diesel (Test 937) had essentially the same horsepower (38.92) as the gasoline model with 11.18 hp hr/gal.

John Deere 2510 Power-Shift Diesel   915
1965-1968
50.66-45.22 hp @ 2500 rpm, 3.86" x 4.33";
202.7 cu in. 4 cyl John Deere engine;
8 forward gears, 1.7 to 17.5 mph;
11.62 hp hr/gal; 6745 lbs.

The gasoline Model (Test 913) with a smaller displacement (180.4 cu in.) and a stroke of 3.86 in. developed 49.57 hp and 7.64 hp hr/gal.

The 2510 Syncro Range Diesel (Test 916) had same displacement and rpm as power-shift, but 54.96 hp and 13.27 hp hr/gal.

The 2510 Syncro Range Gasoline (Test 914) has same displacement and rpm as power-shift model, but developed 53.74 hp and 8.86 hp hr/gal.

FARM TRACTORS 1950-1975

John Deere 5020 Diesel   947
1965-1972
133.25-115.73 hp @ 2200 rpm, 4 3/4" x 5";
531 cu in. 6 cyl John Deere engine;
8 forward gears, 1.9 to 17.9 mph; @ 1900 rpm
13.49 hp hr/gal; 16,045 lbs.

Ford 3000 Diesel   881
1965-1968
39.20-35.71 hp @ 2000 rpm, 4.2" x 4.2";
175 cu in. 3 cyl Ford engine;
8 forward gears, 1.4 to 17.4 mph;
13.25 hp hr/gal; 4141 lbs.

Model 3000 Select-O-Speed Diesel (Test 882) with same displacement and rpm developed 38.06 hp and 11.89 hp hr/gal.

Ford 2000 Gasoline   884
1965-1968
30.57-27.75 hp @ 1900 rpm, 4.2" x 3.8";
158 cu in. 3 cyl Ford engine;
12 forward gears, 2.2 to 14.1 mph;
9.52 hp hr/gal; 3770 lbs.

Ford 4000 Select-O-Speed Diesel   892
1965-1968
45.62-39.43 hp @ 2200 rpm, 4.4" x 4.4";
201 cu in. 3 cyl Ford engine;
10 forward gears, 1.1 to 17.6 mph;
10.69 hp hr/gal; 4895 lbs.

Model 4000 8 Speed Diesel (Test 893) with same displacement and rpm developed 46.71 hp and 12.27 hp hr/gal.

1950-1975 FARM TRACTORS

Ford 5000 Select-O-Speed Diesel 880
1965-1968
54.17-45.98 hp @ 2100 rpm, 4.2" x 4.2";
233 cu in. 4 cyl Ford engine;
10 forward gears, 1.0 to 16.4 mph;
12.48 hp hr/gal; 5825 lbs.

Model 5000 8 Speed Diesel (Test 879) with same displacement and rpm developed 55.96 hp and 12.63 hp hr/gal.

International 4100 Diesel 931
1965-1968
110.82-116.15 hp @ 2400 rpm, 4 1/2" x 4 1/2";
429 cu in. 6 cyl turbocharged International Harvester engine;
8 forward gears, 2.0 to 20.25 mph;
11.52 hp hr/gal; 15,175 lbs.

International Harvester began factory installation of cabs with heater and air conditioner for International-Farmall 1206 and 4100.

Ford Commander 6000 Diesel 878
1961-1967
66.93-63.02 hp @ 2430 rpm, 3.62" x 3.9";
241.7 cu in. 6 cyl Ford engine;
10 forward gears, 1.3 to 19.9 mph;
12.37 hp hr/gal; 7130 lbs.

International Farmall 656 Diesel 912
1965-1973
61.52-54.35 hp @ 1800 rpm, 3 11/16" x 4.39";
281.3 cu in. 6 cyl International Harvester engine;
10 forward gears, 1.7 to 15.9 mph;
12.63 hp hr/gal; 7230 lbs.

Model 656 Gasoline (Test 909) with 262.5 cu in. displacement (3 9/16" x 4.39" bore and stroke) and same rpm developed 63.85 hp and 9.44 hp hr/gal.

FARM TRACTORS 1950-1975

International Farmall 1206 Diesel   910
1965-1967
112.64-99.16 hp @ 2400 rpm, 4 1/8" x 4 1/2";
361 cu in. 6 cyl turbocharged International Harvester engine;
16 forward gears, 1.5 to 19.5 mph;
12.70 hp hr/gal; 10,115 lbs.

First International 2-wheel drive over 100 hp. First Farmall tractor equipped at factory with turbocharger.

Massey-Ferguson 135 Diesel   895
1965-1975
37.82-33.51 hp @ 2000 rpm, 3.6" x 5";
152.7 cu in. 3 cyl Perkins engine;
12 forward gears, 1.3 to 18.8 mph;
14.26 hp hr/gal; 3645 lbs.

Model 135 Gasoline (Test 899) with 4-cylinder 145 cu in. Continental engine and same rpm developed 35.36 hp and 8.94 hp hr/gal.

Kubota RV Gasoline   906
1965-
8.81-6.78 hp @ 2000 rpm, 3.74" x 3.54";
38.5 cu in. 1 cyl Kubota engine;
6 forward gears, 0.6 to 8.2 mph;
6.56 hp hr/gal; 1287 lbs.

Vee belt drive from engine to transmission.

Massey-Ferguson 150 Diesel   901
1965-1975
37.88-33.50 hp @ 2000 rpm, 3.6" x 5";
152.7 cu in. 3 cyl Perkins engine;
12 forward gears, 1.4 to 18.6 mph;
15.02 hp hr/gal; 4805 lbs.

Model 150 Gasoline also available, but not tested.

Massey-Ferguson 165 Diesel   896
1965-1975
52.42-46.94 hp @ 2000 rpm, 3.6'' x 5'';
203.5 cu in. 4 cyl Perkins engine;
12 forward gears, 1.3 to 18.5 mph;
14.74 hp hr/gal; 5085 lbs.

Model 165 Gasoline (Test 898) with 4-cylinder 176 cu in. Continental engine at same rpm developed 46.92 hp and 9.43 hp hr/gal.

Massey-Ferguson 175 Diesel   897
1965-1974
63.34-55.69 hp @ 2000 rpm, 3.875'' x 5'';
236 cu in. 4 cyl Perkins engine;
12 forward gears, 1.3 to 19 mph;
14.22 hp hr/gal; 6125 lbs.

Massey-Ferguson 180 Diesel   900
1965-1974
63.68-54.99 hp @ 2000 rpm, 3.875'' x 5'';
236 cu in. 4 cyl Perkins engine;
12 forward gears, 1.3 to 19.6 mph;
14.78 hp hr/gal; 6755 lbs.

Model 180 Gasoline (Test 1021) with same displacement and rpm developed 62.33 hp and 8.25 hp hr/gal.

Massey-Ferguson 1100 Diesel   923
1965-1972
93.94-85.63 hp @ 2200 rpm, 3.875'' x 5'';
354 cu in. 6 cyl Perkins engine;
12 forward gears, 2.2 to 17.2 mph;
13.19 hp hr/gal; 11,435 lbs.

Model 1100 Gasoline (Test 963) with 320 cu in. displacement and same rpm developed 90.29 hp and 8.02 hp hr/gal.

FARM TRACTORS 1950-1975

Minneapolis-Moline G 1000 LPG 954
1965-1969
110.76-99.18 hp @ 1800 rpm, 4 5/8'' 5'';
504 cu in. 6 cyl Minneapolis-Moline engine;
10 forward gears, 1.8 to 19.4 mph;
7.82 hp hr/gal; 10,475 lbs.

Model G 1000 Diesel (Test 953) with same displacement and rpm developed 110.78 hp and 11.98 hp hr/gal.

Model G 1050 LPG and Diesel versions were manufactured from 1969-1971 with similar performance.

Oliver 1250 Diesel No Test
1965-1969
Certified @ 38.5 hp @ 2500 rpm, 3 5/16'' x 4'';
137.9 cu in. 4 cyl Fiat engine;
6 forward gears, 0 to 15.5 mph.

Equipped with differential lock.

Tractor manufactured by Fiat in Italy was adapted to Oliver specifications.

Oliver 1550 Diesel 943
(Also Minneapolis-Moline G 550)
1965-1969
53.50-45.77 hp @ 2200 rpm, 3 5/8'' x 3 3/4'';
232 cu in. 6 cyl Oliver engine;
12 forward gears, 1.9 to 13.6 mph;
10.10 hp hr/gal; 7020 lbs.

Model 1550 Gasoline (Test 944) with same displacement and rpm developed 53.34 hp and 8.54 hp hr/gal.

Model 1550 LPG tractor had certified 51 hp (No Test).

Oliver 1655 Diesel 1041
1965-1975
70.57-61.90 hp @ 2200 rpm, 3 7/8'' x 4'';
283 cu in. 6 cyl Oliver engine;
18 forward gears, 2.2 to 17 mph;
12.54 hp hr/gal; 8070 lbs.

Same tractor as Minneapolis-Moline G 750.

Model 1655 Gasoline (Test 1042) with 265 cu in. displacement (3 3/4'' x 4'' bore and stroke) and same rpm developed 70.27 hp and 8.12 hp hr/gal.

1950-1975 FARM TRACTORS

1966

Deere with ROPS Appears at Nebraska

The first tractor tested at Nebraska using a Roll-Over Protective frame (ROPS) appeared in the form of a John Deere 4020 LPG tractor (Test 934). Deere released its patents on the Roll-Gard to the entire industry as a move to encourage safety for the tractor operators.

Tractor for Slopes

Steep slopes and hillsides had long caused problems for tractor operators. The danger of a roll-over always existed, whether the operator was a crewman mowing a roadside or a farmer tilling a hilly field.

To meet this problem, Slope Tractor, Inc., Harper, Kansas introduced a leaning wheel tractor which it manufactured until late 1970. Its axles could tilt left or right 30 degrees, remaining parallel to the ground while the tractor's body, engine and wheels remained vertical. The axles could be tilted automatically or manually as desired and could be locked in any position.

Two models of the Slope tractor were the Slope Runner Model 1 and Model 2. The Model 1 weighed 5400 pounds and was equipped with a 4-cylinder Ford industrial engine, either gasoline or diesel. Engines were available in 192 or 172 cu in. displacement and operated at 2000 rpm. The transmission had an 8-speed gear box, giving a range of travel speeds from 1 to 18 mph.

Slope Runner 2 weighed 2620 pounds and was equipped with a Ford 4-cylinder industrial gasoline engine having a 98 cu in. displacement running at 2400 rpm. The transmission was hydrostatic with a hydraulic motor in each rear wheel. Travel speed of this tractor ranged up to 7 mph.

Deutz with Air-Cooled Engine

Many American farmers had never heard of the Deutz tractor when the West German air-cooled diesel was imported into the US. But this company was to become perhaps the largest manufacturer of air-cooled diesel engines in the world.

The Deutz Company had been in business for years. Founded in 1864, it produced its first atmospheric free piston gas engine in 1867 and its first diesel engine in 1898. In 1907 it built what was then known as the "Deutz Motor Plow." In 1924 it introduced a 14 hp diesel tractor.

When it exported its first tractors to the US, Deutz established headquarters for the sale of engines, tractors, and implements in Atlanta with regional sales outlets in Atlanta, Davenport, and Columbus, Ohio. The air-cooled Deutz diesel engines ranged in size from the single cylinder with 8 hp to the large 500-hp units.

Oliver Adds to Its Line

Oliver introduced the 1750 with a 6-cylinder gasoline or diesel engine. The helical transmission provided 12 forward speeds with the use of the hydra-power. Another unusual feature available on the Model 1750 was the over/under hydraul-shift transmission changing the regular 6-speed transmission to provide 18 forward travel speeds—6 speeds 20 percent over and 6 speeds 20 percent under the regular transmission travel speeds.

The new Massey-Ferguson 1130 Diesel was the first Massey-Ferguson tractor to use a turbocharger. The power was increased over 25 hp using the same 354 cu in. displacement as the Model 1100. Because of higher power it was necessary to provide a crankcase oil cooler. A heat exchanger was used in the regular radiator to cool the crankcase oil. This was a common procedure as horsepower continued to rise.

Massey also announced the 130 Diesel. This tractor was very similar to the 25 except for sheet metal changes. Both tractors were built in France.

North Dakota State Highway Dept. safety frame. Early development began in 1959.

1950-1975 FARM TRACTORS

This same year Versatile Manufacturing Company of Winnipeg, Manitoba, entered the 4-wheel drive market. The Versatile tractor was to become popular in Canada and parts of the US.

The new Case 1030 Diesel Comfort King was introduced as a larger size 2-wheel drive tractor in the 100 hp range. This was the first Case tractor to be equipped with a dry type air cleaner. This 1030 was available only with diesel engine and 8-speed transmission, and featured such options as 3-point hitch with Draft-O-Matic and 1000 PTO.

A new transmission plant was built two miles west of Racine, Wisconsin to take care of increased production. New additions were added to the Clausen Plant and also to the foundry.

J.I. Case 1031 Diesel   952
1966-1969
101.79-92.62 hp @ 2000 rpm, 4 3/8'' x 5'';
451 cu in. 6 cyl Case engine;
8 forward gears, 2.0 to 16.2 mph;
12.01 hp hr/gal; 9335 lbs.

First Case tractor tested at Nebraska equipped with a dry type air cleaner.

Massey-Ferguson 130 Diesel   949
1966-1972
26.96-23.06 hp @ 2250 rpm, 3 1/8'' x 3 1/2'';
107 cu in. 4 cyl Perkins engine;
8 forward gears, 1.3 to 16 mph;
11.07 hp hr/gal; 3210 lbs.

Massey-Ferguson 1130 Diesel   948
1966-1972
120.51-109.65 hp @ 2200 rpm, 3.875'' x 5'';
354 cu in. 6 cyl turbocharged Perkins engine;
12 forward gears, 2.1 to 16.7 mph;
13.16 hp hr/gal; 13,560 lbs.

First turbocharged engine used in Massey-Ferguson tractor during Nebraska test.

FARM TRACTORS 1950-1975

Oliver 1750 Diesel 962
1966-1969
80.05-68.09 hp @ 2400 rpm, 3 7/8'' x 4 3/8'';
310 cu in. 6 cyl Oliver engine;
12 forward gears, 1.2 to 14.5 mph;
12.42 hp hr/gal; 9990 lbs.

Model 1750 Gasoline (Test 961) with 283 cu in. displacement (3 7/8'' x 4'' bore and stroke) and same rpm developed 80.31 hp and 8.71 hp hr/gal.

Versatile D-100 Diesel No Test
1966-1967
Manufacturer's estimate of 125 hp @ 2800 rpm, 4.05'' x 3.5'';
363 cu in. 6 cyl Ford engine;
4 forward gears.

Also Model G-100 Gasoline.

First Versatile 4-wheel drive tractor.

1950-1975 FARM TRACTORS

1967

New Hydrostatic Tractor

By this time agricultural tractor manufacturers had been making changes and refinements in transmissions for a number of years. In 1961, the International Harvester Company installed a hydrostatic transmission in its experimental turbine-powered tractor, the HT-340. Now the firm was ready to unveil its hydrostatic tractor, the Model 656.

Basically, the hydrostatic all-speed drive consisted of a piston-type hydraulic pump and a piston-type hydraulic motor. The pump converted mechanical energy from the engine into hydraulic energy; and the motor, connected to the input shaft of a high-low transmission, converted the hydraulic energy back to mechanical energy for power at the drawbar.

The hydraulic pump fastened directly to the flywheel, eliminating the need for an engine clutch. However, a foot pedal was provided to enable the operator to stop the tractor at any time or to maneuver the vehicle in close quarters by releasing the oil pressure.

The final drive components, such as differential, bull gears, and rear axles, were the same as those used on the 656 gear-drive tractors. Travel speeds, ranging up to 8 mph in low range and to 20 mph in high range, were set by moving a single lever, called a speed ratio (S/R) lever located to the left of the steering wheel.

Gave Precise Speed Control

Besides performing the tillage operations of the gear-drive tractors, the hydrostatic tractors were suited particularly to operations where precise travel speed control was required. The job of operating a front loader was easier, because the operator could change directions simply by moving the S/R lever from forward to reverse and back without having to declutch and shift gears. Infinite travel speed was possible within ranges. The hydrostatic tractors were powered by 6-cylinder gasoline or diesel engines.

This same year, International Harvester introduced its "56" line of tractors—the 756, 856, and 1256. Restyling the sheet metal changed the appearance of these tractors over their predecessors. A new factory-installed cab with heater and air conditioner also was introduced this year for these models. The new cab offered more operator room. It was made by Excel Industries, Hesston, Kansas.

International also made a factory installed two-post protective frame with canopy available on the 656, 756, 856 and the 1256 model tractors.

Here's how hydrostatic works: Tractor engine input shaft (1) drives hydrostatic pump (2). When operator moves SR control (3), servo cylinders (4) are actuated, which causes pump swash plate (5) and corresponding swash plate on the motor to tilt. Swash plates tilt but do not rotate. Swash plates are mounted on trunnions (6) inside of transmission case. Pistons (7) in the pump are caused to reciprocate by the sliding action on the tilted swash plate as pump is rotating. The center section (8) separates pump (2) and motor (9) and is also fixed to transmission case. There are open passages in the center section for oil to pass between pump and motor. Oil is the only power connection between pump and motor.

Other Model Changes

Elsewhere in the agricultural tractor industry:
- The Allis-Chalmers Models 170 and 180 replaced the Model D-17, with an improved power train and horsepower increase.
- The David Brown Model 1200 Selectamatic became the first David Brown tractor to use a hand clutch to control the PTO and the hydraulic system.
- The Massey-Ferguson 1080 Diesel tractor was announced.
- J. I. Case was reorganized establishing separate Agri-

cultural, Construction and Components Divisions as the Kern County Land Company (majority Case stockholders) joined with Tenneco Inc., Houston, Texas. This organization is a large controller of gas and oil interests, chemicals, packaging and related investments.
- Finally, the new Case transmission plant was formally opened and dedicated. Full production began that year.
- The newly introduced Oliver 1450 was a Fiat (Italy) built tractor having a 4-cylinder engine. The Oliver 1950-T introduced this year had no connection with the 1950 GM engine series. This tractor had a turbocharged White-Oliver, 6-cylinder diesel engine. It was the first turbocharged Oliver agricultural tractor. Another unusual feature was the use of two fender fuel tanks to supplement the main supply tank. Each fender tank could carry 39 gals.
- The Minneapolis-Moline 900 Series were the first row-crop tractors in the line to approach 100 hp. This series was later known as the G-950 Series and all were available with gasoline, LPG or diesel engines.

The G1000 Vista replaced the earlier Model G1000 Series. Features included an isolated platform mounted on rubber pads for operator comfort and a new location for the fuel tank behind the operator's platform. The LPG and diesel engines were essentially the same as in the G1000 model. No gasoline model was available.

Allis-Chalmers 180 Diesel   964
1967-1970
64.01-56.81 hp @ 2000 rpm, 3 7/8" x 4 1/4";
301 cu in. 6 cyl Allis-Chalmers engine;
8 forward gears, 2.3 to 14.5 mph;
12.17 hp hr/gal; 6510 lbs.

Model 180 Gasoline (Test 1005) with 265 cu in. displacement (3 3/4" x 4" bore and stroke) and same rpm developed 65.16 and 8.91 hp hr/gal.

Allis-Chalmers 170 Diesel   965
1967-1973
54.04-48.62 hp @ 1800 rpm, 3 7/8" x 5";
235.9 cu in. 4 cyl Perkins engine;
8 forward gears, 2.0 to 13.3 mph;
13.94 hp hr/gal; 5950 lbs.

Model 170 Gasoline (Test 966) with 226 cu in. (4" x 4 1/2" bore and stroke) displacement and essentially the same horsepower, developed 9.97 hp hr/gal.

David Brown Selectomatic 1200 Diesel   976
1967-1971
65.23-56.44 hp @ 2300 rpm, 3.94" x 4.5";
219.4 cu in. 4 cyl David Brown engine;
6 forward gears, 2.4 to 14.3 mph;
14.27 hp hr/gal; 6040 lbs.

Hand clutch operated PTO and hydraulic system.

International 444 Gasoline   985
1967-1971
38.09-33.11 hp @ 2000 rpm, 3 3/8" x 4 1/4";
152.1 cu in. 4 cyl International Harvester engine;
8 forward gears, 1.5 to 15.5 mph;
8.47 hp hr/gal; 4130 lbs.

In 1971, four Post ROPS (roll over protection) became available for International Harvester tractors.

International Farmall 706 (also 756) Diesel   955
1967-1971
76.09-68.64 hp @ 2300 rpm, 3.875" x 4.375";
309.6 cu in. 6 cyl International Harvester engine;
16 forward gears, 1.25 to 18.5 mph;
12.47 hp hr/gal; 9160 lbs.

Model 706 LPG (Test 956) with 290.8 cu in. displacement and 2300 rpm developed 76.30 hp and 7.47 hp hr/gal.

Model 706 gasoline (Test 957) with 290.8 cu in. displacement and 2300 rpm developed 76.56 hp and 9.50 hp hr/gal.

International Farmall 656 Hydrostatic Gasoline   968
1967-1973
65.80-50.78 hp @ 2300 rpm, 3 9/16" x 4.39";
263 cu in. 6 cyl International Harvester engine;
Infinitely variable forward speed 0 to 21 mph;
6.89 hp hr/gal; 7215 lbs.

New concept in transmission with infinitely variable hydrostatic drive using a variable displacement hydraulic pump and motor. Sliding gears give high and low ranges.

Model 656 Hydrostatic Diesel (Test 967) with 281.3 cu in. displacement (3 11/16" x 4.39 bore and stroke) developed 66.06 hp and 9.06 hp hr/gal.

International Farmall 856 Diesel   970
1967-1971
100.49-90.09 hp @ 2400 rpm, 4.321" x 4.625";
406.9 cu in. 6 cyl International Harvester engine;
16 forward gears, 1.25 to 18.75 mph;
12.78 hp hr/gal; 10,290 lbs.

1950-1975 FARM TRACTORS

International Farmall 1256 Diesel   971
1967-1969
116.12-104.66 hp @ 2400 rpm, 4.321" x 4.625";
406.9 cu in. 6 cyl turbocharged International Harvester engine;
16 forward gears, 1.5 to 19.5 mph;
12.27 hp hr/gal; 10,405 lbs.

Massey-Ferguson 1080 Diesel   1022
1967-1972
81.23-73.07 hp @ 2000 rpm, 4.5" x 5";
318 cu in. 4 cyl Perkins engine;
12 forward gears, 1.3 to 19.4 mph;
13.84 hp hr/gal; 8520 lbs.

Minneapolis-Moline G 900 LPG   979
(Also Model G 950)
1967-1969
97.57-86.88 hp @ 1800 rpm, 4 1/4" x 5";
425 cu in. 6 cyl Minneapolis-Moline engine;
10 forward gears, 2.2 to 17.4 mph;
7.86 hp hr/gal; 10,160 lbs.

Model G 900 Gasoline (Test 980) with same displacement and rpm developed 97.81 hp and 9.22 hp hr/gal.

Model G 900 Diesel (Test 978) with 451 cu in. displacement (4 3/8" x 5" bore and stroke) and same rpm developed 97.78 hp and 12.04 hp hr/gal. (Minneapolis-Moline G 950 Series very similar).

Minneapolis-Moline G 1000 Vista Diesel   974
1967-1969
111.00-100.97 hp @ 1800 rpm, 4 5/8" x 5";
504 cu in. 6 cyl Minneapolis-Moline engine;
10 forward gears, 2.3 to 18 mph;
11.95 hp hr/gal; 12,230 lbs.

Replaced earlier Model G 1000.

Model G 1000 Vista LPG (Test 975) with same displacement and rpm developed 110.94 hp and 7.78 hp hr/gal.

Introduction of rear-mounted fuel tanks behind the operators seat. Tractor equipped with isolated platform on rubber mounts.

FARM TRACTORS 1950-1975

Oliver 1450 Diesel No Test
1967-1969
Manufacturer's estimate of 55 hp @ 1900 rpm, 4 1/4'' x 4.72'';
268.3 cu in. 4 cyl Fiat engine;
7 forward gears.

Versatile D-118 Diesel No Test
1967-1971
Advertised @ 140 hp @ 3000 rpm, 4 5/8'' x 3.5'';
352 cu in. V6 cyl Cummins engine;
9 forward gears, 3.3 to 20.8 mph; 11,000 lbs.

Oliver 1950 T Diesel 969
1967-1969
105.24-92.74 hp @ 2400 rpm, 3 7/8'' x 4 3/8'';
310 cu in. 6 cyl turbocharged Oliver engine;
18 forward gears, 1.5 to 17.2 mph;
12.72 hp hr/gal; 10,460 lbs.

First Oliver tested at Nebraska using a turbocharger and fender fuel tanks.

Model 1950 T Diesel with 4-wheel drive (Test 972) developed same PTO power displacement and rpm, but slightly less drawbar horsepower (91.30) and 12.52 hp hr/gal.

1968

Test Work on Cab Safety in Sweden

Tractor safety continued to command attention, not only in the US but also in some countries abroad, particularly those with highly developed mechanization in agriculture.

The National Swedish Testing Institute for Agricultural Machinery had noted with concern the rise in tractor accidents and was seeking ways to check them. Among safety options considered by the Institute were various cabs and frames designed to give anti-crush protection to the driver both sideways and rearward in case of overturns.

One of the tests developed and used at the Institute was the pendulum impact test. The pendulum consisted of a welded steel-plate box filled with scrap iron and concrete. The weight was about 2200 lbs. This box was suspended from pivot points by two chains, and could be raised or lowered to suit the height of the tractor cab. The pivot points of the chains were about 20 ft above floor level. The pendulum was pulled back with an electric hoist and released by a quick-release mechanism operated by a rope. Damage by the impact of the pendulum striking the tractor cab then was measured.

After much testing of reinforced cabs and frames, the Institute concluded that tractor cabs will not be completely adequate until they are "designed and manufactured at the same time as the tractors and from the very beginning form an integral part of them."

The Nebraska Tractor Testing Laboratory used the pendulum test on a tractor cab at its Tractor Power and Safety Day in 1968.

Guard for Power Take-Off

But cabs were not the only feature on tractors about which safety engineers were concerned. Farmers continued to suffer mangled hands and feet caught in whirling power take-off shafts. John Deere's answer to this problem was the POWER-GARD for the PTO attachment. Patents for POWER-GARD were released to the industry to help promote safety.

This also was the year that the Northern Manufacturing Company, Havre, Montana came into being and produced its first tractor, the Big Bud. The unusual feature of this tractor was that the engine, transmission, and radiator all were mounted on a skid as an integral unit that could be removed from the tractor for servicing. The company was to produce a limited number of these tractors within the next several years.

International Harvester offered hydrostatic models for the first time in its 544 tractors. The firm also came out with a magnetic pulse transistorized ignition system. With this system, the cam, breaker points, and condenser were no longer needed.

Other Model Changes

Other International Harvester changes, in 544 Series, included larger 4-cylinder engines in gasoline and diesel units. Engine displacement increased from 153 cu in. to 200 cu in. in gasoline units, and from 188 cu in. to 239 cu in. in diesel units. The diesel engine was imported from the firm's Neuss plant in West Germany. Rated engine rpm was 2400 for hydrostatic drives and 2200 for gear drives. Restyling of sheet metal and new decals changed the appearance of the 1968 models.

The new Oliver 2050 and 2150 both used White-designed 6-cylinder Hercules engines operating at 2400 rpm. The two tractors were basically the same, except the 2150 was equipped with a turbocharger. The 2050 had 118 certified hp and the 2150 131 certified hp. Both tractors had planetary gears on axle ends to relieve the load on the drive train in the transmission.

Pendulum test at Nebraska Power and Safety Day. Ruggedness of tractor cab was measured by damage caused by impact of swinging weight.

In 1968, the 1,000th tractor was tested at Nebraska. The operator of the Ford 5000 Diesel was Elmer Schlaphoff, director of Agriculture for the State of Nebraska.

Ford introduced its Model 8000, the first Ford farm tractor with more than 100 PTO horsepower.

Air Cleaners

The oil bath type of air cleaner had been used for many years but the dry type air cleaners were gaining in popularity. The first John Deere tractor tested in 1968 (Test 992) was equipped with a dual type dry air cleaner. From this time on most tractors were converted to the dual type of dry air cleaner.

Two new divisions, Drott Manufacturing of Wausau, Wisconsin and Davis Manufacturing Company of Wichita, Kansas were acquired by Tenneco Inc. and leased to J. I. Case. Case achieved balance of sales between agricultural and construction equipment.

Big Bud HN 250 Diesel   **No Test**
1968-1975
Manufacturers estimate of 310 hp @ 2100 rpm;
855 cu in. 6 cyl Cummins engine;
12 forward gears, 34,000 lbs.

First tractor built by Northern Manufacturing Co., Havre, MT.

FARM TRACTORS 1950-1975

John Deere 820 Diesel — No Test
1968-1970

Manufacturer's estimate of 31 hp @ 2100 rpm, 3.86" x 4.33";
152 cu in. 3 cyl John Deere engine.

Model 830 was identical to 820 announced in 1970 except for sheet metal changes.

John Deere Mannheim Works, West Germany.

John Deere 1520 Diesel — 991
1968-1972

46.52-38.24 hp @ 2500 rpm, 4" x 4.33';
164.9 cu in. 3 cyl John Deere engine;
8 forward gears, 1.5 to 17 mph;
11.96 hp hr/gal; 5855 lbs.

Selective gear fixed ratio operator-controlled partial range power-shifting.

Model 1520 Gasoline (Test 1004) with same displacement and rpm developed 47.86 hp and 8.14 hp hr/gal.

John Deere 2520 Syncro-Range Gasoline — 1003
1968-1972

60.16-55.84 hp @ 2500 rpm, 3.86" x 4.33";
202.7 cu in. 4 cyl John Deere engine;
8 forward gears, 1.8 to 15.8 mph;
8.38 hp hr/gal; 7130 lbs.

Replaced Model 2510.

Model 2520 Syncro-Range Diesel (Test 992) with 219.8 cu in. displacement (4 in. bore) and same rpm developed 61.29 hp and 13.27 hp hr/gal.

Model 2520 Power-Shift Diesel (Test 993) with same displacement and rpm as Test 992, developed 56.28 hp and 10.83 hp hr/gal.

Model 2520 Power-Shift Gasoline (Test 1002) with same displacement and rpm as Test 1003, developed 56.98 hp and 7.15 hp hr/gal.

Deutz D-4006 Diesel — 1075
1968-1979

36.95-32.30 hp @ 2150 rpm, 3.94" x 4.72";
172.5 cu in. 3 cyl air-cooled Deutz engine;
8 forward gears, 1.2 to 15.5 mph;
14.00 hp hr/gal; 4485 lbs; 93.5 dB(A).

Oil cooler coil in air cool blast.

Manufactured by Klockner-Humbold Duetz A.G., Cologne, West Germany.

1950-1975 FARM TRACTORS

Ford 3000 Gasoline 1051
1968-1974
38.33-32.25 hp @ 2100 rpm, 4.2" x 3.8";
158 cu in. 3 cyl Ford engine;
6 forward gears, 1.5 to 18.2 mph;
9.00 hp hr/gal; 4140 lbs.

Model Select-O-Speed (Test 885) with same displacement and rpm developed 36.49 hp and 8.68 hp hr/gal.

Ford 8000 Diesel 973
1968-1972
105.74-89.29 @ 2300 rpm, 4.4" x 4.4";
401 cu in. 6 cyl Ford engine;
8 forward gears, 1.8 to 17.3 mph;
12.69 hp hr/gal; 9980 lbs.

International Farmall 544 Hydrostatic Gasoline 1007
1968-1973
53.87-40.89 hp @ 2400 rpm, 3 13/16" x 4.39";
200.3 cu in. 4 cyl International Harvester engine;
Infinitely variable forward speeds 0 to 21.5 mph;
6.30 hp hr/gal; 6875 lbs.

Model 544 Hydrostatic Diesel (Test 1029) with 239 cu in. displacement (3.875" x 5.06" bore and stroke) and same rpm developed 55.52 hp and 9.29 hp hr/gal.

Also available with conventional gear drive.

Model 544 Diesel (Test 983) 238.6 cu in. 2200 rpm; 52.95-45.44 hp and 12.79 hp hr/gal.

Model 544 Gasoline (Test 984) 200.3 cu in. 2200 rpm, 52.84-45.51 hp and 9.53 hp hr/gal.

Oliver 2150 Diesel 986
1968-1969
131.48-114.63 hp @ 2400 rpm, 4 9/16" x 4 7/8";
478 cu in. 6 cyl turbocharged Oliver engine;
18 forward gears, 1.5 to 16.8 mph;
11.64 hp hr/gal; 12,990 lbs.

Model 2050 Diesel was essentially the same as the 2150 without the turbocharger, developing 118.78 hp (Test 987) and 12.15 hp hr/gal.

FARM TRACTORS 1950-1975

Ursus C-350 Diesel   982
1968
43.34-37.67 hp @ 2000 rpm, 3.74'' x 4.33'';
190.3 cu in. 4 cyl Z M Ursus engine;
10 forward gears, 0.7 to 15.9 mph;
13.12 hp hr/gal, 4880 lbs.

Manufactured at Zaklady Mechaniczne Ursus, Warsaw, Poland.

Model C-335 Diesel (Test 981) with 2-cylinder 119.6 cu in. displacement @ 2200 rpm developed 29.30 hp and 12.06 hp hr/gal.

Model C-325 Diesel (Test 804) with 2-cylinder 110.8 cu in. displacement engine @ 2000 rpm developed 24.60 hp and 13.52 hp hr/gal.

1969

Three new tractor names—Satoh, Leyland and the Case Agri-King—appeared on the American scene.

The Satoh was the product of the Satoh Agricultural Machine Company, Ltd., Tokyo, Japan. The firm established offices in New York City in 1969 and in succeeding years was to introduce several tractors to the American market. Later the company was to merge with the Mitsubishi Agricultural Corporation. The name of the new company was the Mitsubishi Agricultural Machinery Company, Ltd.

The Leyland was a redesigned Nuffield tractor, manufactured at the Leyland Scotland plant in West Lothian. The color of the tractor was changed from orange to a 2-tone blue. PTO horsepower for the 6-cylinder engine was near 100.

Agri-King was the name of a new series of agricultural tractors introduced by Case. They replaced the earlier Case 30 Series and had power shift capabilities. The Model 870 Agri-King had three mechanical ranges and four power shift speeds in each range.

More Changes for J. I. Case

By 1969, Tenneco had acquired additional common stock of J. I. Case increasing its holdings to 99.1 percent.

The new 1470 4-wheel drive Traction King Diesel tractor was introduced. It was the largest agricultural tractor made by Case at the time. The steering system was similar to the Model 1200 using hydrostatic for the front wheels and independent hydraulic power steering for the rear wheels. This permitted crab steering, combined front and rear steering, front steering only, or rear steering only. The hydraulic braking system, unlike the Case 1200, operated a caliper type brake on a disc located on the main drive line.

The 504 cu in. displacement engine in this tractor was the first of the new open-chamber (direct injection) series of Case diesel engines which replaced the Lanova-type, precombustion series engines. A manual transmission with 8 forward and 4 reverse speeds was used.

A new series of 2-wheel drive agricultural tractors, known as the 70-Series Agri-King line was also introduced this year, replacing the earlier "30-Series" models. The four basic models (770, 870, 970, and 1070) incorporated many new features, including rubber-mounted operator platforms for noise and vibration isolation, a choice of three different cushioned operator seats with adjustable suspensions, hydrostatic power steering as standard equipment, and non-metallic rear-mounted fuel tanks, for corrosion-resistance and improved visibility and capacity. The new Case open-chamber diesel engines were available in all models, with gasoline engines also available in all but the 1070. LPG engines were no longer available. The 770 was the last Case tractor tested at Nebraska using gasoline as a fuel.

All models featured a choice of standard 8-speed transmission or a new powershift transmission consisting of four mechanical ranges coupled to three powershift speeds, giving 12 forward and 4 reverse speeds. The powershift portion of the new transmission consisted of a compound planetary gear train and 4 hydraulically-actuated wet disc clutches, providing three forward and one reverse powershift selections. An interlock prevented shifting between forward and reverse until the foot clutch was disengaged. The 870 Agri-King Power Shift Diesel was the first Case powershift tractor tested at Nebraska.

Hydraulically-actuated self-adjusting disc brakes were standard on all models, with hydraulic power-assist available as an option. In addition to single speed PTO options, a dual-speed 540-1000 rpm PTO was also available. The double-ended output shaft of the dual speed unit could be removed, turned around and reinserted to automatically shift the PTO into the correct speed to match the external spline.

This year, the "OLD ABE EAGLE" trade mark, that had been so common to see at Case dealers' business places for many years, was retired. This was replaced by a new and more dynamic Corporate Symbol.

Steiger Moves to Fargo

Steiger moved its operations to Fargo, North Dakota. The first tractor built in its new plant was the Wildcat. It was powered by a 175-hp Caterpillar 3145 V-8 engine. Steiger also produced an 800 Series Tiger, powered by a Cummins V-903 engine with 320 horsepower.

New Oliver Tractors

The Oliver 1250-A, introduced in 1969, was the smallest of the entire line. This tractor was a Fiat built diesel with a 3-cylinder 142.8 cu in. displacement engine operating at 2400 rpm with 38.5 certified horsepower. It replaced the Model 1250.

The White-Oliver 1355 used a Fiat built 4-cylinder diesel which developed 51 certified horsepower. This tractor had 8 forward gears with synchromesh shifting between 3-4 and 7-8 gears. It was equipped with differential lock, parking brake, foot accelerator, drop housing with gear reduction and direct drive oil pump for hydraulics and power steering.

With the introduction of the "55" Series in 1969, the 1250-A diesel was changed to 1255 to conform with the "55" Series. It was basically the same 3-cylinder diesel Fiat engine built and prepared at White's Decatur, Georgia assembly plant according to Oliver specifications. This tractor was also available as a four-wheel drive tractor.

The White-Oliver 1555 had a 6-cylinder engine with a certified 53 hp in gasoline or diesel and 51 hp for LPG. It had 232 cu in. displacement engine running at 2200 rpm. This replaced the earlier 1550.

The White-Oliver 1655 had 70 certified hp when using gasoline or diesel, and 66 hp when using LPG. This tractor available with three auxiliary transmissions—hydra-power drive; over/under hydraul-shift with regular transmission; or creeper drive. Over/under hydraul shift gives 20 percent increase in travel speed or 20 percent below regular 6-speed transmission.

The White-Oliver 1855 used a 310 cu in. turbocharged diesel. One unusual feature found in this tractor was the fuel tanks used on each fender providing 39 gals extra in each fender besides the regular fuel tank located behind the engine. Hydraulic brakes were also used in this new model tractor.

The Oliver 1865 or Minneapolis-Moline G950 was manufactured in the Minneapolis plant and available as a diesel or LPG tractor. This was a 6-cylinder 4 1/4 in. × 5 in. bore and stroke engine developing 97 PTO horsepower. Fender fuel tanks increased the fuel capacity from 40 to 118 gallons.

The Oliver 2155 (same as Minneapolis-Moline G1350 except for paint) developed 135 PTO hp using the Minneapolis-Moline 585 cu in. diesel engine with M.A.N. combustion system. Tractor was also available to use LPG.

The White Plainsman was introduced in 1970 and the Oliver 2655 in 1971. These were versions of the MMA4T 4-wheel drive tractor made in 1969. The A4T was manufactured at White Equipment Co. plants at Minneapolis.

Stress on Economy and New Models

In 1969 the John Deere 4000 Diesel was introduced in an effort to provide a more economical tractor similar to the 4020 minus trimmings. Since the engine was the same as the one used in the 4020 and could be throttled back to lower engine rpm it was reasoned that a user could operate this tractor pulling the same implements as before but travel faster by having more power available than he had before. In this way the additional power would permit more work.

The Model 4520 Diesel was the first John Deere tractor engine equipped with a turbocharger. The lubricating system is almost double the capacity, radiator is larger as well as the fan. For comparison see the following performance of two engines with same displacement:

Test	Model	Disp.	Max PTO hp	Gal/hr	(10 hr) hp hr/gal
1024	John Deere 4020 Diesel	404 cu in.	95.83	6.486	12.50
1014	John Deere 4520 PS Diesel	404 cu in.	122.36	8.113 (Turbo)	12.58

In the above case the horsepower has been increased over 26 hp by using a turbocharger and increasing the fuel supply to the engine. This was the reason that it was necessary to strengthen the entire tractor to take care of the added power.

International Harvester's new models included the 826, 1026, 1456, and 4156. The 826 tractors offered hydrostatic as well as gear-drive transmissions and were available with 6-cylinder gasoline, LPG, or diesel engines. The diesel engine (in the 826) was imported from the firm's Neuss plant in West Germany. The 1456 Model was the first International 2-wheel drive tractor over 130 PTO horsepower.

The 4156 model replaced the 4100 as the firm's large 4-wheel drive tractor. The new model carried the same specifications as the 4100, but the model number was changed to coincide with the 56 Series of 2-wheel drive tractors.

The new Ford Model 9000 turbocharged diesel was the first Ford tractor over 130 hp and the most powerful agricultural tractor Ford had produced so far. It was the first Ford farm tractor to use a turbocharger.

Two Allis-Chalmers Diesels

Two Allis-Chalmers Diesels made their appearance, the 220 and the 160. The 220 developed more than 130 maximum horsepower, which was about 27 hp more than the D-21 which it replaced. The 160, with a maximum horsepower of 40.36, was built by the Renault Tractor Company of France according to Allis-Chalmers specifications. It replaced the Model D-15.

Minneapolis-Moline tractors were now taking on the Oliver appearance more and more. The Minneapolis-Moline G1050 replaced the G1000 Vista.

Cross section of engine used in John Deere Model 4520 diesel tractor, the first John Deere to be equipped with a turbocharger.

The G1350 Series diesel model had a 585 cu in. engine while the LPG engine had 504 cu in. This new series was equipped with hydrostatic power steering. Up to this time all previous Minneapolis-Moline diesels used the Lanova fuel injection system. Beginning with the G1350 series, the diesels used direct injection systems. This tractor series had factory installed cabs available.

Model G955 and G1355 tractors succeeded the G950 and G1350 tractors. These tractors were produced at the White Farm (formerly Oliver) tractor plant in Charles City. The G955 tractor was marketed in Canada as the Model 1870; the G1355 was the Model 2270. These tractors were produced during 1973 and 1974.

Allis-Chalmers 160 Diesel   1028
(Also Model 6040)
1969-1975
40.36-38.07 hp @ 2250 rpm, 3.6" x 5";
152.7 cu in. 3 cyl Perkins engine;
10 forward gears, 0.6 to 15.5 mph;
12.96 hp hr/gal; 4680 lbs.

Built by Renault in France to Allis-Chalmers specifications.

Allis-Chalmers Two-Twenty Diesel   1017
1969-1972
135.95-121.56 hp @ 2200 rpm, 4 1/4" x 5";
426 cu in. 6 cyl turbocharged Allis-Chalmers engine;
8 forward gears, 1.7 to 17.4 mph;
12.21 hp hr/gal; 12,160 lbs.

J.I. Case 770 Power-Shift Gasoline   1032
1969-1972
53.53-47.99 hp @ 1900 rpm, 4" x 5";
251 cu in. 4 cyl Case engine;
12 forward gears, 1.8 to 17 mph;
7.58 hp hr/gal.

Selective gear fixed ratio with partial range operator controlled Power-Shift.

Model 770 Manual Diesel (Test 1033) same rpm, 267 cu in. displacement developed 56.36 hp and 12.92 hp hr/gal.

J.I. Case 870 Power-Shift Diesel   1030
1969-1970
70.53-63.80 hp @ 1900 rpm, 4 5/8" x 5";
336 cu in. 4 cyl Case engine;
12 forward gears, 1.8 to 17 mph;
12.93 hp hr/gal; 9390 lbs.

Model 870 Manual Diesel (Test 1031), with same displacement and rpm developed 70.67 hp and 13.37 hp hr/gal.

First two-wheel drive Case agricultural tractor to use hydrostatic power steering.

1950-1975 FARM TRACTORS

J.I. Case 1470 Traction King Diesel   1006
1969-1972
144.89-132.06 hp @ 2000 rpm, 4 5/8" x 5";
504 cu in. 6 cyl tubocharged Case engine;
8 forward gears, 2.5 to 13.9 mph;
14.39 hp hr/gal; 17,300 lbs.

Hydrostatic steering for front wheels and independent hydraulic steering for rear wheels.

Cast 430 L Diesel   No Test
1969
Manufacturer estimate 32 hp @ 2600 rpm, 3.62" x 3.94";
81.16 cu in. 2 cyl air-cooled Slanzi engine;
6 forward gears, 1 to 13 mph.

Built in Italy—distributed by Reco Sales Inc., Indianapolis, Indiana until late 1970.

Unusual feature of this Italian built 4-wheel drive tractor was that the operator's seat could swing around the steering wheel 180 degrees and lock in place. In this position the driver could operate the tractor forward or reverse and face the direction of travel. The controls were available without any alternation and the transmission provided the same travel speeds for either direction of travel.

John Deere 4000 Diesel   1023
1969-1972
96.89-84.82 hp @ 2200 rpm, 4 1/4" x 4 3/4";
404 cu in. 6 cyl John Deere engine;
8 forward gears, 1.8 to 16.8 mph;
13.13 hp hr/gal, 8605 lbs.

John Deere 4520 Power-Shift Diesel   1014
1969-1972
122.36-111.41 hp @ 2200 rpm, 4 1/4" x 4 3/4";
404 cu in. 6 cyl turbocharged John Deere engine;
8 forward gears, 1.7 to 18.5 mph;
12.58 hp hr/gal; 14,175 lbs.

Precleaner and two dry type filters.

Model 4520 Syncro Range Diesel (Test 1015) had same displacement and rpm developing 123.39 hp and 12.79 hp hr/gal.

FARM TRACTORS 1950-1975

Ford 9000 Diesel   1027
1968-1972
131.22-115.57 hp @ 2200 rpm, 4.4" x 4.4";
401 cu in. 6 cyl turbocharged Ford engine;
16 forward gears, 1.5 to 17.8 mph;
13.0 hp hr/gal; 11.610 lbs.

International 4100 Diesel   931
1965-1970
116.15 drawbar hp @ 2400 rpm, 4 1/2" x 4 1/2";
429 cu in. 6 cyl turbocharged International Harvester engine;
8 forward gears, 2 to 20.25 mph; 15,175 lbs.

International 4156 Diesel same as Model 4100 (Test 931).

International 1026 Hydro Diesel   1047
1969-1976
112.45-83.94 hp @ 2400 rpm, 4.321" x 4.625";
407 cu in. 6 cyl turbocharged International Harvester engine;
Infinitely variable forward speeds, 0 to 17 mph;
9.61 hp hr/gal; 11,100 lbs, 95.5 dB(A).

Hydrostatic drive using a variable displacement pump and motor.

Sliding gears give high and low ranges.

International Farmall 826 Hydro Diesel   1046
1969-1971
84.66-63.54 hp @ 2400 rpm, 3 7/8" x 5.06";
358 cu in. 6 cyl International Harvester engine;
Infinitely variable forward speeds, 0 to 18 mph;
9.64 hp hr/gal; 10,385 lbs, 98.0 dB(A).

Hydrostatic Drive using a variable displacement hydraulic pump and motor. Sliding gears give high and low ranges.

First sound test on International tractors was Test 1045.

1950-1975 FARM TRACTORS

International Farmall 1456 Turbo Diesel   1048
1969-1971
131.80-118.35 hp @ 2400 rpm, 4.321'' x 4.625'';
407 cu in. 6 cyl turbocharged International Harvester engine;
16 forward gears, 1.5 to 20.25 mph;
12.53 hp hr/gal; 12,800 lbs., 92.0 dB(A).

Minneapolis-Moline G-1350 Diesel   1069
1969-1971
141.44-125.29 hp @ 2200 rpm, 4.75'' x 5.50'';
585 cu in. 6 cyl Minneapolis-Moline engine;
10 forward gears, 2.4 to 19.2 mph;
12.23 hp hr/gal; 12,050 lbs., 98.0 dB(A).

Also produced as Oliver 2155 Diesel.

Minneapolis-Moline G-1000 LPG   954
1967-1969
110.76-99.18 hp @ 1800 rpm, 4 5/8'' x 5'';
504 cu in. 6 cyl Minneapolis-Moline engine;
10 forward gears, 1.8 to 19.4 mph;
7.82 hp hr/gal; 10,475 lbs.

Same as Model G-1050 LPG.

Model G-1000 Diesel and G-1050 Diesel (Test 953) with same displacement and rpm developed 110.78 hp and 11.98 hp hr/gal.

Minneapolis-Moline A4T-1600 Diesel   1070
1969-1971
143.27-127.76 hp @ 2200 rpm, 4 3/4'' x 5 1/2'';
585 cu in. 6 cyl Minneapolis-Moline engine;
10 forward gears, 2 to 22.2 mph;
12.24 hp hr/gal; 19,850 lbs., 87.0 dB(A).

Also produced as Oliver 2655 Diesel.

FARM TRACTORS 1950-1975

Oliver 1855 Diesel 1040
1969-1971
98.60-84.30 hp @ 2400 rpm, 3 7/8" x 4 3/8";
310 cu in. 6 cyl turbocharged Oliver engine;
18 forward gears, 1.4 to 18.6 mph;
12.23 hp hr/gal; 11,1140 lbs, 95 dB(A).

Also produced as Minneapolis-Moline G940.

Satoh S605G Gasoline 1106
1969
22.03-18.78 hp @ 2800 rpm, 2.667" x 2.667";
60.2 cu in. 4 cyl Mazda engine;
6 forward gears, 1 to 11.6 mph;
9.03 hp hr/gal; 2240 lbs, 95.5 dB(A).

In 1973 the name changed to "Bison".

Steiger Tiger Diesel No Test
1969-1971
Manufacturer's advertised 320 hp @ 2600 rpm, 5.5" x 4.75";
903 cu in. V8 cyl Cummins engine;
10 forward gears, 2.7 to 17.7 mph; 21,000 lbs.

White-Oliver 1555 Gasoline or Diesel No Test
1969-1971
53 hp (certified by manufacturer) @ 2200 rpm;
232 cu in. 6 cyl Oliver engine;
51 hp for LPG (certified by manufacturer.)

Also produced as White Minneapolis-Moline G550.

1950-1975 FARM TRACTORS

1970

Reducing Tractor Noise

For two decades, evidence had been building that the tractor presented a source of occupational deafness for farmers. Audiograms of farmers who drove tractors tended to prove the harmfulness of tractor noise.

Manufacturers were aware of this problem. In fact, in the early 1940's some manufacturers were attempting to reduce tractor noise by using improved transmission gears.

The Oliver Corporation found that helical gears helped control noise in the gear box. Some of Oliver's models of the 1950's, equipped with helical gears, had much quieter transmissions.

Manufacturers also considered the noise levels of fans, gears, hydraulic pumps, engines, sheet metal and mufflers, to determine whether these devices could be made less noisy.

But the tractor cab appeared to offer the greatest promise for reducing noise, although earlier cabs were thought to increase rather than decrease noise.

No Uniform Measures

One of the problems which noise researchers encountered was a lack of uniformity in measurements. Since various reporting scales were used, it was difficult to compare noise levels of tractors prior to 1970. One unit of measurement was the decibel, which measured sound pressure. During the year it became apparent the dBA scale would be accepted universally. The U.S. Department of Labor requested this scale be used in its legal procedures.

The Nebraska Tractor Testing Laboratory began sound tests on all tractors it tested and in December, 1970 noise test results were published in the dBA scale.

Standardized sound test procedures gave the farmer a basis on which he could compare sound levels of various tractors. These procedures also provided manufacturers with a new sales incentive for reducing noise. Within the next several years, improvements in cabs alone had reduced tractor noise nearly one-half.

New Case Tractors

During these years the agricultural tractor industry was stirring with name changes, transfers of stock, consolidations, and various other business transactions. Shareholders of the J. I. Case Company approved a plan making the company a wholly-owned subsidiary of Tenneco, Inc., Houston, Texas.

TABLE 1. NEBRASKA TRACTOR TEST SOUND LEVEL DATA (1970-1978)

Year	Number of tractors tested Without cabs	With cabs	Total	Average sound levels, dB(A)* Without cabs	With cabs	All Tractors
1970	22	7	29	95.4	93.7	95.0
1971	8	19	27	96.9	89.1	91.4
1972	14	14	28	96.2	86.3	91.3
1973	18	19	37	96.7	85.0	90.7
1974	3	11	14	96.2	84.0	86.6
1975	13	17	30	95.9	87.2	91.0
1976	17	12	29	95.5	83.0	90.3
1977	7	28	35	97.3	80.3	88.7
1978	11	19	30	92.6	80.9	85.2

*Average sound levels were determined by mathematically averaging the sound levels of all tractors in each group.

From: Schnieder, 1978.

Table 1 shows the results of noise tests for each year since 1970. The dB(A) scale is such that a reduction of 10 dB(A) is equivalent to decreasing the effect on the human ear by one-half. Therefore, a sound of 90 dB(A) is twice as loud as one of 80 dB(A) and four times as loud as a sound of 70 dB(A).

TABLE 2. EXPOSURE LEVEL. OSHA NOISE LIMITS PER DAY

Sound level dB(A)	Time permitted hours-minutes	Sound level dB(A)	Time permitted hours-minutes
85	16-0	101	1-44
86	13-56	102	1-31
87	12-8	103	1-19
88	10-34	104	1-9
89	9-11	105	1-0
90	8-0	106	0-52
91	6-56	107	0-46
92	6-4	108	0-40
93	5-17	109	0-34
94	4-36	110	0-30
95	4-0	111	0-26
96	3-29	112	0-23
97	3-2	113	0-20
98	2-50	114	0-17
99	2-15	115	0-15
100	2-0	116+	None

From: Schnieder, 1978.

Table 2 lists the maximum safe exposure time recommended for the various sound levels dB(A) by Occupational Safety and Health Administration.

Case extended the new family of 4- and 6- cylinder in-line open-chamber diesel engines to construction and forestry equipment. The new engines ranged in size from 65 to 180 horsepower.

The Case 1090 and 1170 2-wheel drive tractors were introduced. The 1090 was the same 100 PTO hp power level as the 1070, but was designed with certain special and heavy-duty features for more rugged usage. These features included heavier drawbar, 3-point hitch, and front axle components, power-assisted caliper disc brakes, and a dual-element dry air cleaner.

The Model 1170 was the largest Case 2-wheel drive tractor built to date, at 121.93 PTO horsepower. In addition to heavy-duty features similar to the 1090, it was equipped with a 451 cu in. displacement turbocharged 6-cylinder engine, 8-speed manual transmission, and outboard planetary final drives.

The first unofficial Nebraska sound tests were made on a Case 970 Manual Diesel tractor (Test 1034). During the 1970 Test season preliminary sound measurements were made on 29 tractors. This provided experience and procedures that later became a part of all future tractor sound tests at Nebraska beginning officially in 1971.

Allis-Chalmers added the Crop Hustler 175 and 185, and the Land Handler 210 Diesel models to its tractor line. Later, the Model 175 was equipped with a 248 cu in. Perkins engine, replacing the former 236 cu in. engine. The Model 210 was similar to the Model 220, with about 14 less horsepower.

Oliver engineering was moved from Charles City to Hopkins, Minnesota to further consolidate White Engineering for the combined companies. The Oliver 2655 4-wheel drive (same as the Minneapolis-Moline A4T-1600) was introduced in 1971. This model tractor was the first Oliver tractor tested at Nebraska with a factory installed cab.

The Kubota Iron and Machinery Works, Ltd., Japan, shortened its name to Kubota, Ltd. The company also opened an engineering office in California. Its 20 hp diesel was now selling in the US.

The Massey-Ferguson 1150 was the firm's first tractor equipped with a V-8 Perkins diesel.

Case Model 970 was first tractor tested for sound at Nebraska Tractor Testing Laboratory in 1970. Sound-measuring instruments are carried in vehicle towed along side the test car.

Allis-Chalmers 175 Gasoline 1156
1970-1976
60.88-53.17 hp @ 1800 rpm, 4' x 4 1/2'';
226 cu in. 4 cyl Allis-Chalmers engine;
8 forward gears, 2 to 13.1 mph;
9.64 hp hr/gal; 6390 lbs, 91.0 dB(A).

Early use of roll-over protection.

Allis-Chalmers Crop Hustler 175 Diesel 1043
1970-1978
62.47-55.27 hp @ 2000 rpm, 3 7/8'' x 5'';
236 cu in. 4 cyl Perkins engine;
8 forward gears, 2.1 to 13.7 mph;
13.33 hp hr/gal; 5925 lbs, 93 dB(A).

Allis-Chalmers Crop Hustler 185 Diesel 1044
1970-1978
74.87-65.07 hp @ 2200 rpm, 3 7/8" x 4 1/4";
301 cu in. 6 cyl Allis-Chalmers engine;
8 forward gears, 2 to 16 mph;
11.89 hp hr/gal; 6710 lbs, 96 dB(A).

Replaced Model 180 with higher engine speed.

Big Bud HN 320 Diesel No Test
1970-1978
Manufacturer advertised 320 hp @ 2100 rpm, 5.5" x 6.0";
855 cu in. 6 cyl turbocharged Cummins engine
with intercooler;
12 forward gears, 1.6 to 13.6 mph; 36,000 lbs.

Replaced Model 250 with tilting cab. Partial Range power-shift.

Allis-Chalmers Landhandler Two-Ten Diesel 1065
1970-1973
122.40-106.35 hp @ 2200 rpm, 4 1/4" x 5";
426 cu in. 6 cyl turbocharged Allis-Chalmers engine;
8 forward gears, 1.5 to 16.25 mph;
11.85 hp hr/gal; 12,425 lbs, 97.5 dB(A).

J.I. Case 770 Manual Gasoline 1061
1970
56.32-50.09 hp @ 1900 rpm, 4" x 5";
251 cu in. 4 cyl Case engine;
8 forward gears, 1.9 to 15 mph;
8.17 hp hr/gal; 8810 lbs, 89.0 dB(A).

Model 770 Power-Shift Diesel (Test 1058) same rpm, 267 cu in. displacement developed 56.77 hp and 12.58 hp hr/gal. 94 dB(A).

1950-1975 FARM TRACTORS

J.I. Case 870 Power Shift Diesel   1076
1971-1973
77.92-67.47 hp @ 2000 rpm, 4 5/8" x 5";
336 cu in. 4 cyl Case engine;
12 forward gears, 1.8 to 17 mph;
12.41 hp hr/gal; 10,230 lbs, 88.5 dB(A).

Selective gear fixed ratio with partial range operator controlled power-shifting.

Model 870 Manual Diesel (Test 1077) with same displacement and rpm developed 77.45 hp and 13.44 hp hr/gal. 88.5 dB(A).

J.I. Case 970 Power Shift Diesel   1037
1969-1970
85.31-74.48 hp @ 1900 rpm, 4 1/8" x 5";
401 cu in. 6 cyl Case engine;
12 forward gears, 1.8 to 17 mph;
12.85 hp hr/gal; 10,335 lbs, 95.0 dB(A).

Model 970 Manual Diesel (Test 1034) with same displacement and rpm developed 85.70 hp and 12.93 hp hr/gal. 94.5 dB(A)

Model 970 Power-Shift Gasoline (Test 1038) with 377 cu in. displacement and same rpm developed 85.23 hp and 8.15 hp hr/gal. 91.5 dB(A)

Model 970 Manual Gasoline (Test 1039) with 377 cu in. displacement and same rpm developed 85.02 hp and 8.40 hp hr/gal. 92.0 dB(A)

The first official sound tests were made on a J.I. Case 970 Manual Diesel (Test 1034) at Nebraska, 94.5 dB(A).

J.I. Case 1070 Manual Diesel   1067
1969-1970
107.36-91.45 hp @ 2100 rpm, 4 3/8" x 5";
451 cu in. 6 cyl Case engine;
8 forward gears, 2 to 15.8 mph;
12.62 hp hr/gal; 11,410 lbs, 89.0 dB(A).

Model 1070 Power-Shift Diesel (Test 1066) with same displacement and rpm developed 108.10 hp and 12.68 hp hr/gal. 92.5 dB(A)

J.I. Case 1170 Diesel 1062
1970-1971
121.93-110.46 hp @ 2100 rpm, 4 3/8'' x 5'';
451 cu in. 6 cyl turbocharged Case engine;
8 forward gears, 1.9 to 15 mph;
13.09 hp hr/gal; 13,540 lbs, 91.0 dB(A).

First J.I. Case tractor tested at Nebraska with safety cab and first Case two-wheel drive tractor with turbocharger. Later called Model 1175 Diesel (1971-1978).

Kubota L-210 Diesel 1071
1971-1974
19.10-14.69 hp @ 2700 rpm, 3.46'' x 3.46'';
65.3 cu in. 2 cyl Kubota engine;
6 forward gears, 1.0 to 10.7 mph;
10.56 hp hr/gal; 2320 lbs, 95.5 dB(A).

John Deere 4320 Syncro Range Diesel 1050
1970-1972
116.55-104.40 hp @ 2200 rpm, 4 1/4'' x 4 3/4'';
404 cu in. 6 cyl turbocharged John Deere engine;
8 forward gears, 2 to 18.9 mph;
11.77 hp hr/gal; 10,675 lbs, 92.5 dB(A).

Kubota L-260 Diesel 1072
1971-1976
24.11-19.64 hp @ 2600 rpm, 3.54'' x 3.94'';
77.6 cu in. 2 cyl Kubota engine;
8 forward gears, 0.9 to 8.8 mph;
10.90 hp hr/gal; 2640 lbs, 97.5 dB(A).

Massey-Ferguson 1150 Diesel   1049
1970-1972
135.60-125.63 hp @ 2200 rpm, 4 1/4'' x 4 1/2'';
510.7 cu in. V8 cyl Perkins engine;
12 forward gears, 2.1 to 16.8 mph;
13.37 hp hr/gal; 13,590 lbs, 98.5 dB(A).

Steiger Bearcat Diesel   1079
1970-1975
165.64 db hp @ 2800 rpm, 4.5'' x 5'';
636 cu in. V8 cyl Caterpillar engine;
10 forward gears, 2.4 to 19.3 mph;
12.46 hp hr/gal; 19,560 lbs, 90.5 dB(A).

Crankcase oil cooled by engine coolant.

FARM TRACTORS 1950-1975

1971

More Power with Intercoolers

While the turbocharger proved to be a practical device for adding power and efficiency to an engine, it had its drawbacks.

The problem was that in compressing intake air, the turbocharger also increased air temperature considerably. Tests at the Nebraska Tractor Testing Laboratory indicated that for every 10 degrees rise in air temperature, horsepower was reduced 1 to 2 percent. A device was needed that could cool the air that had been compressed and heated by the turbocharger so that the air would support more fuel and thus increase horsepower. The answer was an intercooler, which acted as a heat exchanger to reduce the temperature of the incoming air to more nearly the temperature of the engine coolant.

Three new John Deere tractors were introduced: the 4620 Diesel, the 7020 4-wheel drive Diesel and the 6030 Diesel. All three new model tractors were equipped with turbochargers and intercoolers. The 7020 model was the first tractor tested at Nebraska using an *intercooler* and turbocharger (Test 1063)

The following table gives a direct comparison of the three tractors with engines having the same displacement and same rpm. One is equipped with a turbocharger and the other two are equipped with turbochargers and intercoolers.

Test	Model	Disp.	Max. PTO hp	Gal/hr
1050	John Deere 4320 SR Diesel	404	116.55	7.953 (Turbocharged)
1073	John Deere 4620 SR Diesel	404	135.76	9.235 (Turbocharged with Intercooler)
1063	John Deere 7020 FWD Diesel	404	146.17	10.236 (Turbocharged with Intercooler)

New Look for International

International Harvester introduced 11 new models; 8 in the 66 Series, plus the 354, 454 and 574. The tractors in the 66 line ranged from 80 to 145 PTO horsepower. Included in the series was Model 1468, International's first farm tractor powered by a V-8 engine with a displacement of 550 cu in., operating at 2600 rpm. 16-speed

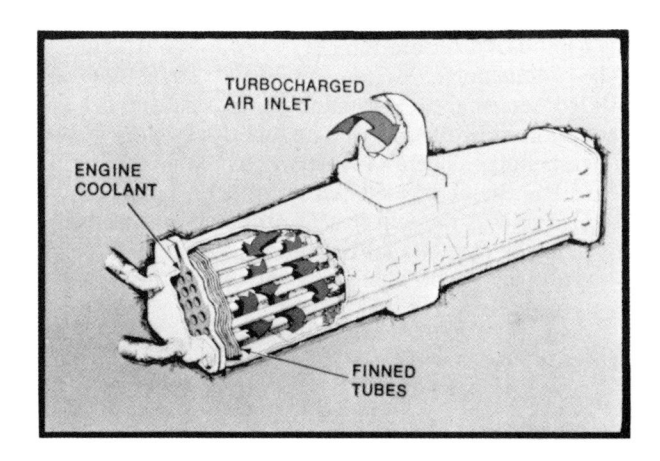

This intercooler was used on the Allis-Chalmers Model 7050 tractor. Located in the intake manifold, it was cooled by the engine coolant system and could lower incoming air from the turbocharger about 100 degrees.

transmissions were available in all models. Other features of the 66 line included:

- Heavily upholstered deluxe cabs with built-in protective frame.
- Restyled sheet metal.
- Differential locks in the 766, 966, and 1066 models.

The International Harvester 354 models offered either gasoline or diesel engines, both having a 144 cu in. displacement operating at a rated speed of 1875 rpm which provided 32 PTO horsepower. These models were imported from International Harvester of Great Britain.

The International Harvester 454 and 574 were diesels powered by 3-cylinder engines. Major features were:

- Newly designed power train featuring 4-speed synchronized constant-mesh transmission, permitting on-the-go shifting.
- Transmission shift levers on left side of seat, hitch and hydraulic levers on right side, clearing area between operator's knees.
- Wet, oil-bathed, hydraulic brakes.
- Planetary units driving rear axles.
- Differential locks.
- Pressure lubricated power train.

More New Models

Ford introduced its Model 7000, with a turbocharged diesel. This tractor also featured a load monitor, a hydraulic draft control system that sensed drive line loads to improve implement control.

Massey-Ferguson's first 4-wheel drive tractor was the 1500, powered by a Caterpillar V-8 diesel. The engine speed of 3000 rpm was the highest rpm for any tractor produced by the firm. This tractor also was equipped with articulated power steering.

Minneapolis-Moline Diesels G 350 and G 450 were Fiat-built tractors. They also were known as the White 1270 and White 1370.

David Brown Tractors, Ltd. featured the "hydra-shift" in its new 1212 Diesel. The automatic transmission provided on-the-move clutchless speed changes to any of four ratios in each of three working ranges. By this time, David Brown was using safety cabs on its tractors.

The Allis-Chalmers Company changed its name to the Allis-Chalmers Corporation.

The Minneapolis-Moline A4T-1400 and 1600 articulated steering tractors were produced at the Lake Street plant in Minneapolis. The A4T-1400 had a 504 cu in. diesel engine. The A4T-1600 was available with a 504 cu in. LPG engine or a 585 cu in. diesel. These tractors were also sold as the Oliver 2455 and 2655, respectively.

In 1971 the first official Nebraska Tests were made on all tractors for sound along with the usual tests.

David Brown 995 Diesel   1136
1971-
58.77-50.42 hp @ 2200 rpm, 3.94'' x 4.5'';
219 cu in. 4 cyl David Brown engine;
12 forward gears, 1.1 to 16.1 mph;
13.6 hp hr/gal; 5150 lbs, 96.0 dB(A).

First David Brown with power steering at Nebraska.

David Brown 885 Diesel   1137
1971-
43.20-37.16 hp @ 2200 rpm, 3.94'' x 4.5'';
164.4 cu in. 3 cyl David Brown engine;
12 forward gears, 1.1 to 16.4 mph;
13.73 hp hr/gal; 4355 lbs, 94.5 dB(A).

David Brown 1210 Diesel   1142
1971
65.98-57.04 hp @ 2300 rpm, 3.94'' x 4.5'';
219 cu in. 4 cyl David Brown engine;
12 forward gears, 1.2 to 17.8 mph;
14.36 hp hr/gal; 5800 lbs, 98.0 dB(A).

FARM TRACTORS 1950-1975

David Brown 1212 Diesel   1138
1971-
65.36-53.88 hp @ 2300 rpm, 3.94" x 4.5";
219 cu in. 4 cyl David Brown engine;
12 forward gears, 1 to 16 mph;
13.27 hp hr/gal; 6405 lbs, 97.5 dB(A).

J.I. Case 2470 Diesel   1114
1971-1978
174.20-154.24 hp @ 2200 rpm, 4 5/8" x 5";
504 cu in. 6 cyl turbocharged Case engine;
12 forward gears, 2 to 15 mph;
12.55 hp hr/gal; 20,485 lbs, 85.0 dB(A).

Transmission was selective gear fixed ratio with partial range operator controlled power shifting. Front and rear steering linkage could be interconnected by means of a single lever so that all wheels could be steered simultaneously.

J.I. Case 970 Manual Diesel   1078
1971-1978
93.87-81.27 hp @ 2000 rpm, 4 1/8" x 5";
401 cu in. 6 cyl Case engine;
8 forward gears, 2 to 15.8 mph;
13.26 hp hr/gal; 11,080 lbs, 88.5 dB(A).

Model 970 Power-Shift Diesel (Test 1068) with same engine rpm and same displacement developed 93.41 hp and 12.37 hp hr/gal. 88.5 dB(A).

John Deere 2030 Diesel   1085
1971-1975
60.65-52.33 hp @ 2500 rpm, 4" x 4.33";
219 cu in. 4 cyl John Deere engine;
16 forward gears, 1.1 to 17.8 mph;
11.82 hp hr/gal; 5370 lbs, 97.0 dB(A).

Model 2030 Gasoline (Test 1084) with same displacement and rpm as the 2030 Diesel developed 60.34 hp and 7.66 hp hr/gal. 94.5 dB(A)

Dubuque tractors.

1950-1975 FARM TRACTORS

John Deere 4620 Syncro Range Diesel   1073
1971-1972
135.76-117.38 hp @ 2200 rpm, 4 1/4" x 4 3/4";
404 cu in. 6 cyl turbocharged John Deere engine
with intercooler;
8 forward gears, 2 to 18.3 mph;
11.86 hp hr/gal; 14,800 lbs, 93.0 dB(A).

Model 4620 Power Shift Diesel (Test 1064) with the same displacement and rpm developed 135.62 hp and 11.27 hp hr/gal. 95 dB(A)

John Deere 6030 Diesel   1100
1971
175.99-155.42 hp @ 2100 rpm, 4.75" x 5";
531 cu in. 6 cyl turbocharged John Deere engine
with intercooler;
8 forward gears, 2 to 18.5 mph;
12.72 hp hr/gal; 18,180 lbs, 86.5 dB(A).

John Deere 7020 Diesel   1063
1971-1974
146.17-131.50 hp @ 2200 rpm, 4 1/4" x 4 3/4";
404 cu in. 6 cyl turbocharged John Deere engine
with intercooler;
16 forward gears, 2 to 21.4 mph;
11.81 hp hr/gal; 18,495 lbs, 96.0 dB(A).

First use of intercooler for John Deere tractors tested at Nebraska.

Deutz D5506 Diesel   1104
1971-1976
56.08-48.78 hp @ 2300 rpm, 3.94" x 4.72";
230 cu in. 4 cyl air-cooled Deutz engine;
8 forward gears, 1.4 to 16.2 mph;
13.69 hp hr/gal; 5290 lbs, 95.5 dB(A).

Transmission synchronized, in all gears except 1st and 2nd.

FARM TRACTORS 1950-1975

Deutz D8006 Diesel   1074
1971-1978
85.51-73.04 hp @ 2100 rpm, 3.94'' x 4.72'';
345 cu in. 6 cyl Deutz air-cooled engine;
16 forward gears, 0.6 to 17.9 mph;
14.17 hp hr/gal; 6850 lbs, 98.0 dB(A).

Similar type transmission as Model D5506.

Manufactured by Klockner-Humboldt Deutz A.G., Cologne, West Germany.

International 354 Gasoline   1115
1971-1973
32.58-28.37 hp @ 1875 rpm, 3 7/8;; x 4'';
143.1 cu in. 4 cyl International Harvester engine;
8 forward gears, 1.5 to 14 mph;
8.55 hp hr/gal; 4035 lbs, 91.0 dB(A).

Imported from International Harvester Ltd., United Kingdom.

International 454 Gasoline   1096
1971-1973
40.86-35.24 hp @ 2200 rpm, 3 3/8'' x 4.39'';
157 cu in. 4 cyl International Harvester engine;
8 forward gears, 2 to 20.25 mph;
8.40 hp hr/gal; 4400 lbs, 92.0 dB(A).

First tractor to use lead free gasoline at Nebrasks test.

Model 454 Diesel (Test 1097) with 179 cu in. 3-cylinder engine (3 7/8'' x 5.06'' bore and stroke) and same rpm developed 40.47 hp and 11.85 hp hr/gal. 94.5 dB(A).

International 574 Utility Diesel   1099
1971-1973
52.55-46.95 hp @ 2200 rpm, 3 7/8'' x 5.06'';
238.6 cu in. 4 cyl International Harvester engine;
8 forward gears, 2 to 21.25 mph;
12.38 hp hr/gal; 5150 lbs, 96.5 dB(A).

Model 574 is the same as International Row Crop Diesel.

Model 574 Utility Gasoline (Test 1098) with 200.3 cu in. displacement (3 13/16'' x 4.39'' bore and stroke) and same rpm developed 52.97 hp and 9.05 hp hr/gal. 94.5 dB(A).

Also International 574 Row Crop.

1950-1975 FARM TRACTORS

International Farmall 766 Diesel   1117
1971-1976
85.45-74.36 hp @ 2600 rpm, 3 7/8'' x 5.085'';
360 cu in. 6 cyl International Harvester engine;
16 forward gears, 1.5 to 19.75 mph;
11.05 hp hr/gal; 11,340 lbs, 88.5 dB(A).

International Farmall 966 Hydro Diesel   1095
1971-1974
91.38-71.33 hp @ 2400 rpm, 4.3'' x 4.75'';
414 cu in. 6 cyl International Harvester engine;
Infinitely variable travel speed 0 to 18 mph; 9.19 hp hr/gal;
11,700 lbs, 90.0 dB(A).

International Farmall 966 Diesel   1123
1971-1976
100.80-86.09 hp @ 2600 rpm, 4.3'' x 4.75'';
414 cu in. 6 cyl International Harvester engine;
16 forward gears, 1.5 to 20.5 mph;
11.68 hp hr/gal; 11,700 lbs, 89.0 dB(A).

"ISO" mounted safety cabs on Models 766, 966, 1066, 1466, and 1468.

International Farmall 1066 Turbo Diesel   1124
1971-1976
125.68-108.73 hp @ 2600 rpm, 4.3'' x 4.75'';
414 cu in. 6 cyl turbocharged International Harvester engine;
16 forward gears, 1.5 to 22.25 mph;
11.97 hp hr/gal; 13,315 lbs, 88.5 dB(A).

FARM TRACTORS 1950-1975

International Farmall 1466 Turbo Diesel   1080
1971-1976
133.40-116.97 hp @ 2400 rpm, 4.3" x 5";
436 cu in. 6 cyl turbocharged International Harvester engine;
16 forward gears, 1.5 to 20.25 mph;
12.85 hp hr/gal; 13,480 lbs, 87.0 dB(A).

International Farmall 1466 Turbo Diesel   1125
1973-1976
145.77-126.91 hp @ 2600 rpm, 4.3" x 5";
436 cu in. 6 cyl turbocharged International Harvester engine;
16 forward gears, 1.5 to 22.25 mph;
12.01 hp hr/gal; 13,670 lbs, 93.5 dB(A).

International Farmall 1468 V-8 Diesel   1118
1971-1974
145.49-128.72 hp @ 2600 rpm, 4 1/2" x 4 5/16";
550 cu in. V8 cyl International Harvester engine;
16 forward gears, 1.5 to 22 mph;
13.22 hp hr/gal; 13,460 lbs, 92.5 dB(A).

First International agricultural tractor using a V-8 engine.

Massey-Ferguson 1500 Diesel   1086
1971-1975
152.77 db hp @ 3000 rpm, 4.5" x 4.5";
573 cu in. V8 cyl Caterpillar engine;
12 forward gears, 2.3 to 21.3 mph;
12.49 hp hr/gal; 17,200 lbs, 88.5 dB(A).

First Massey-Ferguson tractor with 4-wheel drive since WWII.

Massey-Ferguson 1800 Diesel   1087
1971-1975
178.70 db hp @ 2800 rpm, 4.5" x 5";
636 cu in. V8 cyl Caterpillar engine;
12 forward gears, 2.1 to 19.8 mph;
13.49 hp hr/gal; 17,330 lbs, 89.5 dB(A).

Minneapolis-Moline G-350 Diesel   No Test
1971-
Manufacturer estimate 43 hp @ 2400 rpm, 3.94" x 4.33";
158.1 cu in. 3 cyl Fiat engine;
6 forward gears; 3710 lbs.

Also produced as Oliver 1265 Diesel (1971) and White 1270 Diesel (1973).

Manufactured by Fiat S.p.A., Turin, Italy.

Minneapolis-Moline G-450 Diesel   No Test
1971-
Manufacturer estimate 59 hp @ 2400 rpm, 3.93" x 4.33';
210.9 cu in. 4 cyl Fiat engine;
8 forward gears.

Also produced as Oliver 1365 Diesel (1971-) and White 1370 Diesel (1973).

Manufactured by Fiat S.p.A., Turin, Italy.

1972

More New Companies

New tractor companies, some with foreign ties, continued to appear on the American agricultural scene.

One such firm was the Long Manufacturing Company, Tarboro, North Carolina. Shortly after WW II, this company had manufactured tobacco harvesting and drying equipment. Later it added balers, backhoes, mowers, rotary hoes, and a peanut combine to its line. It actively entered the tractor market when it began importing tractors from Italy, Poland, and Romania and "Americanizing" them for US buyers.

Long set up an assembly line to convert the imports for the domestic market. American wheels, tires, batteries, mufflers, wheel weights, cabs, and other accessories were added. Each tractor received a new paint job. Long sold five basic models, ranging from 32 to 98 horsepower.

About this time a new tractor company sprang up at Great Falls, Montana. It was D. L. Curtis, Inc., dedicated to building a tractor suitable for the large wheat acreages of the Northwest but still easy to service and maintain. The company built one basic model, the 404, with variations which would accommodate different power outputs and tire sizes. The axles and transmissions were designed to handle the large horsepower engines.

J. I. Case added the David Brown Tractor line to its products when Tenneco Inc. acquired the English firm of David Brown Tractors Ltd. and made it an operating Division of the J. I. Case Company.

Several new Case tractors were introduced in 1972 including the 1270 and 1370 Agri-King Diesels. These 2-wheel drive tractors were equipped with turbochargers to increase the power capability. The transmissions were newly designed larger versions of the Case power shift transmission first introduced in the 770 through 1070 models. Inboard planetary final drives, power assisted wet disc brakes, and adjustable front axle with hydrostatic power steering were standard equipment. Available options included 3-point hitch with Draft-O-Matic, 1000 rpm PTO, hydraulically actuated differential lock, a certified ROPS and sound-insulated cab with heater and air conditioning options, and a hydraulic system providing up to four remote circuits and flow controls.

The Model 1175 replaced the 1170 this year. The 1175 contained several improvements, but power level was not changed.

The largest Case tractor built up to this time, the 174 PTO hp 2470 Traction King 4-wheel drive, replaced the Model 1470. This model was equipped with a 504 cu in., turbocharged diesel engine, a newly-designed version of the Case power shift transmission, adjustable wheel tread settings, and an exhaust-aspirated dual element dry air cleaner. The 4-wheel steering system, improved over the systems used on the previous 1200 and 1470 models, could now also be actuated in the 4-wheel steering mode for tight turns by the steering wheel alone.

New Hefty Model F

The Hefty Tractor Company, Juneau, Wisconsin, a division of the Holtan Axle and Transmission Company, Inc., began marketing tractors on a national basis in 1972 with its introduction of the Hefty Model F. The tractor was intended to serve as a replacement for the Ford 8N tractor, but foreign competition forced reduction of domestic sales efforts.

The Steiger Company put the Cougar on the market with a 250-hp Caterpillar 6-cylinder engine having a 638 cu in. displacement. The tractor was one of the first Steiger units with a turbocharger to boost power. The tractor had 10 speeds, ranging from 1.6 to 14.1 mph and weighed 21,000 lbs.

Oliver/Minneapolis-Moline Tractors

Oliver unveiled its Model 2255, the first Oliver tractor to be equipped with a 573 cu in. Caterpillar V-8 diesel engine. This tractor was designed as a 2-wheel drive vehicle, with an optional mechanical-powered front wheel drive attachment.

Oliver with 18 Travel Speeds

The Minneapolis-Moline G1355 was introduced at the same time as the Oliver 2255. Production of the Minneapolis-Moline tractors had been transferred from the Lake Street plant in Minneapolis to the Charles City plant. The Model G1355 tractor was available with the Minneapolis-Moline 585 cu in. diesel engine or the 504 cu in. LPG engine. The Oliver-developed Over/Under Hydraul-shift was combined with the 6-speed helical gear transmission, which provided operator controlled power shift and 18 forward travel speeds. Most of the production appeared in the Minneapolis-Moline paint colors. The new model was also produced with the White trade name as a Model 2270 for the Canadian market.

Oliver also marketed its Models 1265 and 1365. Built by Fiat in Italy, these tractors were given a final assembly in the US with tires, sheet metal, and electrical systems added. The 1265 used a 3-cylinder diesel engine that produced 47 horsepower. The 1365 used a 4-cylinder diesel engine that produced 63 horsepower, according to Oliver sources.

The first large 4-wheel drive tractor sold by Allis-Chalmers was the Model 440 Diesel. Built by Steiger in Fargo, it incorporated Allis-Chalmers styling and innovations. Only a limited number were sold.

The Allis-Chalmers Model 200 Diesel replaced the Model 190XT. The 200 had new styling and a Comfort Cab.

For Diesel Fuel Only

Many of John Deere's Generation II tractors were equipped to handle only diesel fuel. These included Models 4030, 4230, 4430, and 4630. A new style cab adorned the line. The company identified the cab as the "Sound-Guard." The frame consisted of a 4-post structure that provided roll-over protection. The new design also reduced sound level in the cab. New transmissions offered the operator a choice between an 8-speed Syncro-Range, a 16-speed Quad-Range, and Power Shift.

The new Massey-Ferguson Models, 1085, 1105, 1135, and 1155, were equipped with tilting cabs to provide easier access to the tractor when servicing or repairing.

The International Harvester Company's new models included the 4166, 666, and 664 tractors. The 4166 and 666 tractors featured larger displacement engines. Engine speed on the 666 tractors was increased from 1800 to 2000 rpm on gear drive models and from 2300 to 2400 rpm on models with hydrostatic drive. Drawbar horsepower on the 4166 was 130. The color of the 4166 was changed to red and white, compared to the yellow and white of its predecessor. International offered the 666 tractors in Farmall models only.

The Model 664 was a utility tractor in the 60-65 PTO hp class. Its engine was a 4-cylinder diesel with a displacement of 239 cu in. operating at 2400 rpm.

Allis-Chalmers 200 Diesel   **No Test**
1972-1974
93.64-84.06 hp @ 2200 rpm, 3 7/8" x 4 1/4";
301 cu in. 6 cyl turbocharged Allis-Chalmers engine;
8 forward gears, 2.1 to 13.6 mph;
13.07 hp hr/gal; 7945 lbs.

Sold on basis of Test 887 performed on Model One-Ninety XT. All data taken from this test.

Massey-Ferguson Models 1155, 1135, and 1105 featured improved 3-point linkage as standard equipment on row-crop models in 1972. Larger hydraulic cylinders on the 1155 and 1105 tractors increased lift capacity by about 20 percent. Rub blocks (circled), replacing check chains on lower links, could be adjusted for stabilization through entire lift range or in transport position only.

Allis-Chalmers 440 Diesel   **No Test**
1972-1975
Manufacturer's estimate of 208-165 hp @ 2800 rpm, 4 5/8" x 4 1/8";
555 cu in. V8 cyl Cummins engine;
10 forward gears, 2.3 to 18.8 mph;
17,500 lbs.

Manufactured by Steiger Tractor Inc., Fargo, ND.

J.I. Case 1270 Diesel   1103
1972-1973
126.70-108.28 hp @ 2100 rpm, 4 3/8" x 5";
451 cu in. 6 cyl turbocharged Case engine;
12 forward gears, 1.9 to 17.8 mph;
11.55 hp hr/gal; 13,440 lbs, 88.5 dB(A).

John Deere 4030 Quad-Range Diesel   1111
1972
80.33-68.25 hp @ 2500 rpm, 4" x 4.33";
329 cu in. 6 cyl John Deere engine;
16 forward gears, 2.0 to 16.4 mph;
10.99 hp hr/gal; 9180 lbs, 83.5 dB(A).

J.I. Case 1370 Diesel   1102
1972-1973
142.51-123.01 hp @ 2100 rpm, 4 5/8" x 5";
504 cu in. 6 cyl turbocharged Case engine;
12 forward gears, 1.9 to 17.8 mph;
11.82 hp hr/gal; 15,290 lbs, 89.0 dB(A).

John Deere 4230 Quad-Range Diesel   1112
1972
100.32-86.36 hp @ 2200 rpm, 4 1/4" x 4 3/4";
404 cu in. 6 cyl John Deere engine;
16 forward gears, 1.9 to 16.7 mph;
11.92 hp hr/gal; 10,900 lbs, 82.5 dB(A).

1950-1975 FARM TRACTORS

John Deere 4430 Quad-Range Diesel   1110
1972
125.88-108.46 hp @ 2200 rpm, 4 1/4" x 4 3/4";
404 cu in. 6 cyl turbocharged John Deere engine;
16 forward gears, 2 to 17.8 mph;
11.96 hp hr/gal; 11,350 lbs, 82.5 dB(A).

John Deere 7520 Diesel   1101
1972-1974
175.82-165.21 hp @ 2100 rpm, 4 3/4" x 5";
531 cu in. 6 cyl turbocharged John Deere engine
with intercooler;
16 forward gears, 2.1 to 22.4 mph;
12.82 hp hr/gal; 22,320 lbs, 87.5 dB(A).

John Deere 4630 Power-Shift Diesel   1113
1972
150.66-131.69 hp @ 2200 rpm, 4 1/4" x 4 3/4";
404 cu in. 6 cyl turbocharged John Deere engine
with intercooler;
8 forward gears, 1.7 to 17.7 mph;
12.88 hp hr/gal; 16,250 lbs, 83 dB(A).

Deutz D13006 Diesel   1130
1972-1978
125.77-110.24 hp @ 2400 rpm, 3.94" x 4.72";
345 cu in. 6 cyl turbocharged Deutz air cooled engine;
16 forward gears, 0.6 to 18.8 mph;
13.83 hp hr/gal; 12,050 lbs, 89.0 dB(A).

All transmission gears synchronized except 1st and 2nd.

Crankcase oil cooled by coil in air cooling blast. First Deutz tractor tested with dual tires. This tractor equipped with US built ROPS cab.

Hefty Model F Gasoline   No Test
1972-
Advertised @ 27 hp @ 3300 rpm, 2.28" x 3.14";
51.6 cu in. 4 cyl Continental engine;
6 forward gears, 0.8 to 14 mph;
3000 lbs.

International 666 Diesel   1151
1972-1976
66.29-58.34 hp @ 2000 rpm, 3 7/8" x 4.41";
312 cu in. 6 cyl International Harvester engine;
10 forward gears, 1.8 to 16.5 mph;
11.98 hp hr/gal; 8200 lbs, 94.5 dB(A).

Model 666 Gasoline (Test 1152) 290.8 cu in. displacement (3 3/4" x 4.39" bore and stroke) and same rpm developed 66.30 hp and 9.04 hp hr/gal. 91.5 dB(A)

International 4166 Diesel   1116
1972-1976
150.63-130.47 hp @ 2400 rpm, 4.3" x 5";
436 cu in. 6 cyl turbocharged International Harvester engine;
8 forward gears, 2.0 to 20.5 mph;
11.88 hp hr/gal; 16,325 lbs, 89.0 dB(A).

Color changed to red and white from original yellow.

Long U-445 Diesel   1108
1972
41.93-37.55 hp @ 2400 rpm, 3.74" x 4.33";
143 cu in. 3 cyl Uzina engine;
6 forward gears, 1.2 to 12.9 mph;
13.29 hp hr/gal; 4480 lbs, 98.5 dB(A).

Also manufactured as UTB U-445 Diesel.

Produced by Uzina Tractorul Brasov (UTB), Romania.

1950-1975 FARM TRACTORS

Long U-550 Diesel   1109
1972
53.61-47.88 hp @ 2400 rpm, 3.74'' x 4.33'';
190.7 cu in. 4 cyl Uzina engine;
8 forward gears, 1.8 to 18.4 mph;
13.21 hp hr/gal; 4570 lbs, 96.0 dB(A).

Also manufactured as UTB U-550 Diesel.

Produced by Uzina Tractorul Brasov (UTB), Romania.

Long R9500 Special Diesel   1107
1972
97.72-82.94 hp @ 2250 rpm, 3.875'' x 5.0'';
354 cu in. 6 cyl Perkins engine;
12 forward gears, 1.1 to 15.3 mph;
13.10 hp hr/gal; 8170 lbs, 99.0 dB(A).

Manufactured by Massey-Ferguson-Landini S.p.A. Fabrico, Reggio Emilia, Italy.

Massey-Ferguson 1105 Diesel   1132
1972-1977
100.72-86.26 hp @ 2200 rpm, 3.875'' x 5'';
354 cu in. 6 cyl turbocharged Perkins engine;
12 forward gears, 2.2 to 16.7 mph;
11.55 hp hr/gal; 12,320 lbs, 82.5 dB(A).

One of the first Massey-Ferguson tractors equipped with a turbocharger.

Massey-Ferguson 1135 Diesel   1135
1972-1977
120.84-106.52 hp @ 2200 rpm, 3.875'' x 5'';
354 cu in. 6 cyl turbocharged Perkins engine;
12 forward gears, 2.2 to 16.9 mph;
12.56 hp hr/gal; 13,550 lbs, 82.5 dB(A).

FARM TRACTORS 1950-1975

Massey-Ferguson 1155 Diesel   1134
1972-1977
140.97-124.90 hp @ 2200 rpm, 4.25" x 5";
540 cu in. V8 cyl Perkins engine;
12 forward gears, 2.2 to 16.7 mph;
13.00 hp hr/gal; 14,906 lbs, 84.5 dB(A).

Rite 303 Diesel   No Test
1972-
Manufacturer advertised 350 hp @ 2100 rpm, 5 1/2" x 6";
855 cu in. 6 cyl Cummins engine;
10 forward gears, 2 to 16.7 mph.

Rite 404 Diesel   No Test
1972-
Manufacturer advertised 450 hp @ 2100 rpm,
6 1/4" x 6 1/4";
1150 cu in. 6 cyl turbocharged Cummins engine;
10 forward gears, 2.0 to 16.7 mph.

Model 505 Diesel is similar to the Model 404 except an intercooler was added boosting the engine to 525 hp (manufacturer's rating).

Model 606 Diesel is similar to the Model 505 except additional fuel flow boosted the engine to 600 hp (manufacturer's rating).

Steiger Cougar Diesel   No Test
1972-1974
250 advertized hp @ 2200 rpm, 4 3/4" x 6";
638 cu in. 6 cyl turbocharged Caterpillar engine;
10 forward gears, 1.6 to 14.1 mph.

1950-1975 FARM TRACTORS

1973

Soviet Belarus Causes Stir

In the early 1970's, the Soviet-built Belarus tractor caused quite a stir at tractor shows and demonstrations across the country. Farmers and engineers alike were curious to see what this Russian tractor looked like and how it performed.

The tractor differed from American agricultural tractors in several ways. It carried an air compressor and pressure tank. The reason was that Soviet law required air brakes on trailers pulled behind tractors. The compressed air also proved convenient for inflating tires and cleaning seed boxes, etc.

The Soviet tractor used a centrifugal-type oil filter instead of the replaceable filter common to American tractors. A bowl within the filter was spun at high speeds by jets of oil. The bowl could be removed easily, cleaned, and replaced.

Coil springs in the front axle spindles of this tractor provided a cushion for the front end assembly when traveling over rough terrain.

The tractor's rear or final drive was equipped with a differential lock that engaged automatically when one drive wheel slipped over 6 percent more than the opposite drive wheel.

Three Belarus tractors with engine horsepower ranging from 29 to 57 used air-cooled diesel engines. The larger models, from 57 to 85 hp, used liquid-cooled diesel engines, and were available with optional front wheel drive attachments. These attachments provided for front wheel drive automatically whenever the rear wheels slipped between 7 percent and 8 percent.

Tractor Seated Two

All the Belarus tractors with 57 to 85 horsepower are available with front-wheel drive. Those over 85 hp are 4-wheel drive with articulated power steering. They also featured cabs with seats wide enough for two persons. Belarus Machinery, Inc. was to come into being within a few years to serve as an agency for marketing the Soviet tractors in the US. Offices were established in New Orleans and Milwaukee.

The Allis-Chalmers Corporation chose the year 1973 to announce a new family of large tractors. Included were the Models 7030 and 7050. Following were some of their features:

- Multi-disc wet master clutches with Power Director shifting capability.
- Independent 2-speed PTO.
- 20 forward travel speeds.
- Hydraulically operated differential lock.
- Planetary final drives.
- Wet multi-disc brakes operated hydraulically.
- All new sheet metal components.

Both the 7030 and the 7050 used turbochargers and the 7050 also used an intercooler, the first Allis-Chalmers tractor to be so equipped. This intercooler, located in the intake manifold, was cooled by the engine coolant system and could lower incoming air from the turbocharger about 100 degrees.

Another innovation on the 7030 and 7050 tractors was swing-out batteries for easier servicing. Normally only two 12-volt batteries were used, but four batteries—two additional ones—were available as optional equipment for cold weather starting. New snap-on panels provided easy access to each side of the engine compartment.

The Ford Tractor Division expanded its range of farm tractors by introducing the Model 1000 and Model 9600. The 1000 was a 23-hp tractor and the 9600 was a 135-hp turbocharged diesel. The 9600 was Ford's largest and most powerful farm tractor to date.

Among John Deere's new models were the 830, 1530, 2030, and 2630, in the 35 to 70 hp range. The John Deere 830 and the 1530 both had 3-cylinder engines and the 2030 and 2630 had 4-cylinder engines. All four tractors were designed to use planetary final drives, differential locks, self-adjusting hydraulic wet disc brakes and power steering.

International Harvester entered the articulated 4-wheel drive tractor field with its Model 4366. This tractor was assembled by Steiger in which International had acquired a vested interest the previous year. The Model 4366 had a turbocharged 225-hp, 6-cylinder diesel engine operating at 2600 rpm.

A Fuller 5-speed constant-mesh transmission with high-low selection provided 10 forward and two reverse speeds. Power was transmitted from the transmission to the axle drive lines through a swinging power divider designed by Steiger. Steiger also designed the front and rear frames, 3-point hitch, cab with protective frame, heater, air conditioner, and hydraulic system.

A caliper-type disc brake, located on the drive line and operated hydraulically by a single foot pedal, provided 4-wheel braking.

Other new International Harvester models this year included the 1566 and 1568 V-8, with PTO horsepower of 160 and 150, respectively. These tractors featured a newly designed 3-speed, constant-mesh transmission followed by a Hi-Lo range section, with optional Torque-Amplifier providing 12 forward and 6 reverse travel speeds.

The International Hydro 100 tractor replaced the 966 and 1066 hydrostatic tractors, and the 464 and 674 utility tractors made their appearance. The gasoline version of the 464 featured a 4-cylinder engine with 175 cu in. displacement, engine rpm of 2400 and PTO horsepower of 44.4. The 674 model was in the 60 PTO hp class. It was available with 4-cylinder gasoline or diesel engines.

One historical note: International discontinued use of the name "FARMALL," a name familiar to farmers and agriculturalists since the 1920's.

Rome a Powerful Tractor

The Rome tractor, built by the Woods-Copeland Company, Wharton, Texas, went on the market in 1973 after three years of testing in the rice fields of Texas. Several years later Woods-Copeland was to be purchased by Rome Industries, a subsidiary of the Wyman-Gordon Company, and moved to Cedartown, Georgia. Future tractors to be produced by this company were to offer tremendous horsepower—450, 475, and 600.

The Rome 600C 4-wheel drive agricultural tractor featured a Cummins KTA-1150C diesel engine rated at 600 hp at 2100 rpm with Allison powershift transmission and Cotta 2-speed dropbox.

Standard equipment on these monoliths included:
- Planetary type, Caterpillar axles.
- Allison powershift transmissions with 12 forward speeds and two reverses.
- Electric over hydraulic implement controls.
- Pressurized cabs.
- Heavy-duty 22,000 BTU air conditioners.
- Audio-visual warning systems.
- AM-FM radio-stereo tape player.

The Steiger Company introduced three new powerful tractors—the Super Wildcat, the Bearcat, and the Tiger.

The Super Wildcat used a 200-hp Caterpillar engine with 573 cu in. displacement. The Bearcat had a 225-hp Caterpillar V-8 engine with 636 cu in. displacement, and the Tiger was equipped with a 320-hp Cummins engine with 903 cu in. displacement.

The Minneapolis-Moline G955 was introduced in 1973. This tractor was a combination of a Minneapolis-Moline engine and an Oliver designed transmission and auxiliary Over/Under transmission. This model tractor was available with either the Minneapolis-Moline 451 cu in. diesel engine or the 425 cu in. LPG engine. In Canada it was sold as the White 1870.

Horsepower of the Case 870 Diesel tractors and the Model 1370 tractor was increase to 80 PTO horsepower and 155 PTO horsepower, respectively. Production of the Model 770 tractor was discontinued. All Case agricultural tractors were now powered by diesel engines.

Allis-Chalmers 7030 Diesel   1119
1973
130.98-113.18 hp @ 2300 rpm, 4 1/4" x 5";
426 cu in. 6 cyl turbocharged Allis-Chalmers engine;
20 forward gears, 1.5 to 19 mph;
11.90 hp hr/gal; 12,430 lbs, 79.5 dB(A).

Factory installed cab.

Allis-Chalmers 7050 Diesel 1120
1973-1974
156.49-135.11 hp @ 2300 rpm, 4 1/4" x 5";
426 cu in. 6 cyl turbocharged Allis-Chalmers engine with intercooler;
20 forward gears, 1.6 to 19.8 mph;
12.67 hp hr/gal; 14,525 lbs, 80.5 dB(A).

Big Bud H-N 350 Diesel No Test
1973-
Advertised @ 350 hp @ 2100 rpm, 5.5" x 6";
855 cu in. 6 cyl turbocharged Cummins engine with intercooler;
13 forward gears, 1.3 to 16.6 mph;
40,000 lbs, advertised 75 dB(A).

Belarus MTZ 80 Diesel 1139
1973
74.79-65.64 hp @ 2200 rpm, 4.331" x 4.921";
288.7 cu in. 4 cyl MTZ engine;
18 forward gears, 1.2 to 23.9 mph;
13.12 hp hr/gal; 7840 lbs, 86.5 dB(A).

Built in USSR at Minsk Tractor Plant.

John Deere 1530 Diesel 1147
1973-1975
45.38-38.86 hp @ 2500 rpm, 4" x 4.33";
164 cu in. 3 cyl John Deere engine;
16 forwad gears, 1.2 to 17.3 mph;
11.47 hp hr/gal; 4954 lbs, 98.5 dB(A).

Built by John Deere Werke, Mannheim, West Germany.

1950-1975 FARM TRACTORS

John Deere 2630 Diesel   1157
1973-1975
70.37-58.15 hp @ 2500 rpm, 4.2" x 5.0";
276 cu in. 4 cyl John Deere engine;
16 forward gears, 1.3 to 16.3 mph;
10.98 hp hr/gal; 5800 lbs, 97.0 dB(A).

Manufactured in Dubuque.

Ford 8600 Diesel   1121
1973
110.69-90.60 hp @ 2300 rpm, 4.4" x 4.4";
401 cu in. 6 cyl Ford engine;
16 forward gears, 1.6 to 18.6 mph;
12.74 hp hr/gal; 11,430 lbs, 82.0 dB(A).

International Hydro 100 Diesel   1158
1973-1976
104.17-82.13 hp @ 2400 rpm, 4.3" x 5.0";
436 cu in. 6 cyl International Harvester engine;
Infinitely variable travel speed 0 to 17 mph;
10.19 hp hr/gal; 12,130 lbs, 91.5 dB(A).

International 464 Gasoline   1126
1973-1976
45.74-42.20 hp @ 2400 rpm, 3.562 x 4.39";
175 cu in. 4 cyl International Harvester engine;
8 forward gears, 2 to 22.75 mph;
8.86 hp hr/gal; 4895 lbs, 95.0 dB(A).

Model 464 Diesel (Test 1127) 3-cylinder engine with 179 cu in. displacement (3.875" x 5.06" bore and stroke) and same rpm developed 44.42 hp and 12.57 hp hr/gal. 98.5 dB(A)

FARM TRACTORS 1950-1975

International 1568 Diesel   1175
1973-1976
150.70-133.79 hp @ 2600 rpm, 4.5" x 4 5/16";
550 cu in. V8 cyl International Harvester engine;
12 forward gears, 1.9 to 17.6 mph;
12.59 hp hr/gal; 14,950 lbs, 90.0 dB(A).

Model 1566 Turbo Diesel (Test 1174) with 6-cylinder, 436 cu in. displacement engine (4.3" x 5.0" bore and stroke) and same rpm developed 161.01 hp and 11.69 hp hr/gal. 88.5 dB(A)

Kubota L225 Diesel   1145
1973-
20.86-17.69 hp @ 2700 rpm, 3.0" x 3.2";
68 cu in. 3 cyl Kubota engine;
8 forward gears, 1.1 to 13.1 mph;
10.78 hp hr/gal; 1995 lbs, 93.0 dB(A).

Also available in 4-wheel drive (Model L225DT).

International 4366 Diesel   1153
1973-1976
Manufacturers estimate of 225-168.90 hp @ 2600 rpm, 4.30" x 5.35";
466 cu in. 6 cyl turbocharged International Harvester engine;
10 forward gears, 2.5 to 20.5 mph;
11.98 hp hr/gal; 21,690 lbs, 86.0 dB(A).

Built by Steiger Tractor Inc. to International Harvester specifications.

Massey-Ferguson 1085 Diesel   1133
1973-1979
81.58-70.76 hp @ 2000 rpm, 4.5" x 5";
318 cu in. 4 cyl Perkins engine;
12 forward gears, 1.4 to 20.3 mph;
13.15 hp hr/gal; 11,540 lbs, 83.5 dB(A).

1950-1975 FARM TRACTORS

Minneapolis-Moline G-955 Diesel   1161
1973-1975
98.38-83.40 hp @ 1800 rpm, 4.375" x 5.0";
451 cu in. 6 cyl Minneapolis-Moline engine;
18 forward gears, 1.5 to 17.0 mph;
11.76 hp hr/gal; 12,060 lbs, 95.0 dB(A).

Replaced Model 950 Diesel. Also manufactured as White 1870 Diesel (1973).

Minneapolis-Moline G-1355 Diesel   1141
1973-1974
142.62-128.55 hp @ 2200 rpm, 4 3/4" x 5 1/2";
585 cu in. 6 cyl Mineapolis-Moline engine;
18 forward gears, 1.5 to 16.9 mph;
12.85 hp hr/gal; 17,096 lbs, 87.0 dB(A).

Replaced Model G-1350.

Also manufactured as White 2270 Diesel (1973).

Equipped with over-and-under Hydraul-shift.

Model G-1355 Diesel was one of the first Minneapolis-Moline's to have cab mounted at factory.

Steiger Super Wildcat   No Test
1973-1975
Manufacturer's estimate of 185 hp @ 3000 rpm, 4.5" x 4.5";
573 cu in. V8 cyl Caterpillar engine;
10 forward gears, 2.4 to 19.7 mph;
15,500 lbs.

Woods and Copeland 450C   No Test
1973
Advertised @ 450 hp, turbocharged Cummins engine with intercooler;
12 forward gears, 2.5 to 24.8 mph;
34,500 lbs.

Hydrostatic articulated steering.

FARM TRACTORS 1950-1975

1974

Influx of 4-Wheel Drives

An influx of large 4-wheel drive models marked this period in the history of agricultural tractors. John Deere introduced two such tractors—the 8430 and the 8630. The latter tractor was similar to the 8430 but with 50 additional horsepower. Both were diesels.

Steiger announced its new Series 11 tractors. Company spokesmen claimed Steiger's new factory was the largest 4-wheel drive tractor plant in the world.

The Oliver 2255 Diesel tractor received a Caterpillar 636 cu in. engine in place of the Cat 573 cu in. engine.

The name Oliver was being changed to WHITE. The first of a complete new line of tractors with the silver and charcoal paint colors bearing the WHITE name appeared in 1974. The White Field Boss 4-150 was the first such tractor. This model was a 4-wheel drive with articulated power steering. The tractor was equipped with the 636 cu in. Caterpillar diesel engine; 4-wheel drive and 150 horsepower, thus the Model 4-150. The Model series number designating the number of drive wheels followed by the PTO horsepower.

An unusual feature of the 4-150 was the low profile design. The engine crankshaft was placed in line with or at the same height as the transmission input shaft. The low profile greatly improved visibility and aided the engine serviceability.

Late in the year White introduced three new 2-wheel drive tractors. The White 2-70 with a 70 hp gasoline engine was the last gasoline model produced. The gasoline engine had 265 cu in. A diesel with 283 cu in. was available as a direct injection engine.

The White 2-105 with 105 horsepower was available as a diesel tractor only, having a 354 cu in. turbocharged engine. The White 2-150 produced 147 horsepower. It was powered by the Minneapolis-Moline 585 cu in. M.A.N. type diesel engine.

Features Rear Engine Design

The Hefty Model G, featuring a rear engine design, appeared on the market as a replacement of the Allis-Chalmers Model G. The Hefty Model G offered more power, additional travel speeds, live PTO, adjustable wheel spacing and variable ground clearance. The Hefty tractors were designed to serve the commercial vegetable farmer, nursery owners, and tobacco growers. A flexible and telescoping rear axle drive shaft made possible the adjustable wheel spacing.

Renault, the French manufacturer, produced the Model 6040 for Allis-Chalmers, but the dollar devaluation soon forced sales of the diesel model to be discontinued in the US.

Three Allis-Chalmers Models

Allis-Chalmers now produced three new models, the 7040, 7060, and 7080. The 7040 and 7060 tractors were equipped with turbochargers. The 7060 also had an intercooler.

The Allis-Chalmers Model 7080 featured an engine that ran at 2550 rpm (250 rpm higher), providing an additional 20 horsepower. This was the largest 2-wheel drive tractor produced by Allis-Chalmers up to this time.

Near Crosby, North Dakota a father and son farming operation prompted development of a new tractor. When Jerome Knudson and his father could not find a tractor which met their wheat-farming needs, the younger Knudson decided to build one himself. The result was the Knudson tractor. This tractor's self-leveling feature made it popular among wheat farmers of the Northwest. Knudson tractors had large comfortable, noise-resistant cabs. Later, Knudson production and sales rights were assigned to Allmand Brothers, Inc., Holdrege, Nebraska.

A new Case 2670 Traction King and an improved 1270 Diesel Agri-King were both introduced. The 2670 was the first Case tractor to have both a turbocharger and an intercooler as part of the engine. This and other changes allowed the tractor to produce over 219 hp at the PTO. Equipment and options available were similar to the model 2470. The Model 870 tractor was discontinued this year. Case changed the color of their tractors from Flambeau Red and Desert Sand to Power Red and Power White on all agricultural equipment division and David Brown equipment. The David Brown tractors had been white, red, and brown.

Allis-Chalmers 6040 Diesel   No Test
1974-1976
Manufacturer's estimate of 40.36-36.71 hp @ 2250 rpm;
152 cu in. 3 cyl Perkins engine;
10 forward gears.

This tractor built to Allis-Chalmers specifications by Renault.

Allis-Chalmers 7040 Diesel   1166
1974
136.49-117.12 hp @ 2300 rpm, 4 1/4" x 5";
426 cu in. 6 cyl turbocharged Allis-Chalmers engine;
20 forward gears, 1.6 to 18.8 mph;
11.86 hp hr/gal; 11,795 lbs, 78.5 dB(A).

Allis-Chalmers 7000 Power-Shift Diesel   1195
1974
106.44-92.64 hp @ 2200 rpm, 3 7/8" x 4 1/4";
301 cu in. 6 cyl turbocharged Allis-Chalmers engine;
12 forward gears, 1.9 to 19.3 mph;
12.27 hp hr/gal; 9550 lbs, 79.0 dB(A).

Allis-Chalmers 7060 Diesel   1197
1974
161.42-144.13 hp @ 2300 rpm, 4 1/4" x 5";
426 cu in. 6 cyl turbocharged Allis-Chalmers engine with intercooler;
12 forward gears, 1.9 to 19.1 mph;
12.98 hp hr/gal; 14,645 lbs, 80.0 dB(A).

FARM TRACTORS **1950-1975**

Allis-Chalmers 7080 Diesel   1168
1974-1975
153.78-181.51 hp @ 2550 rpm, 4 1/4" x 5";
426 cu in. 6 cyl Allis-Chalmers turbocharged engine
with intercooler;
20 forward gears, 1.6 to 18.3 mph;
12.11 hp hr/gal; 14,605 lbs; 83.0 dB(A).

J.I. Case 2670 Diesel   1165
1974-1978
219.44-193.65 hp @ 2200 rpm, 4 5/8" x 5";
504 cu in. 6 cyl turbocharged Case engine with intercooler;
12 forward gears, 2.0 to 14.5 mph;
12.60 hp hr/gal; 20,810 lbs; 83.5 dB(A).

First Case agricultural tractor tested at Nebraska equipped with intercooler.

J.I. Case 1270 Diesel   1160
1974-1978
135.39-120.85 hp @ 2100 rpm, 4 3/8" x 5";
451 cu in. 6 cyl turbocharged Case engine;
12 forward gears, 1.9 to 17.7 mph;
11.49 hp hr/gal; 15,470 lbs; 83.0 dB(A).

John Deere 8430 Diesel   1179
1974
178.16-160.39 hp @ 2100 rpm, 4.563" x 4.75";
466 cu in. 6 cyl turbocharged John Deere engine
with intercooler;
16 forward gears, 2.0 to 20.3 mph;
12.64 hp hr/gal; 22,810 lbs; 80.5 dB(A).

1950-1975 FARM TRACTORS

John Deere 8630 1180
1974
225.59-201.96 hp @ 2100 rpm, 5 1/8" x 5";
619 cu in. 6 cyl turbocharged John Deere engine
with intercooler;
16 forward gears, 2.0 to 20.3 mph;
12.80 hp hr/gal; 24,530 lbs, 82.5 dB(A).

Hefty Model G Diesel No Test
1974-
Advertised @ 27 hp @ 2650 rpm, 3.57" x 3.89";
75.04 cu in. 2 cyl Mitsubishi engine;
6 forward gears, 0.8 to 12 mph;
3000 lbs.

Also available with a Continental gasoline engine.

International Hydro 70 Diesel 1155
1974-1976
69.51-53.49 hp @ 2400 rpm, 3 7/8" x 4.41";
312 cu in. 6 cyl International Harvester engine;
Infinitely variable travel speed 0 to 21.5 mph, 8.75 hp hr/gal;
8440 lbs, 97.0 dB(A).

Also built as Model 666 Diesel and Hydro 666 Diesel.

Knudson 360 Hillside Diesel No Test
1974
Advertised 310 hp @ 2100 rpm, 5 1/2" x 6";
855 cu in. 6 cyl tubocharged Cummins engine
with intercooler;
12 forward gears, 1.3 to 19.1 mph;
24,000 lbs.

Long 900 Diesel 1163
1974
72.88-62.31 hp @ 2200 rpm, 4.33" x 4.72";
285 cu in. 4 cyl Zetor engine;
16 forward gears, 1.1 to 15.4 mph;
13.00 hp hr/gal; 8200 lbs, 100.0 dB(A).

Steiger Bearcat II Diesel 1162
1974
180.69 db hp @ 2800 rpm, 4.5" x 5";
636 cu in. V8 cyl Caterpillar engine;
10 forward gears, 2.0 to 16.1 mph;
12.37 hp hr/gal; 21,435 lbs, 86.5 dB(A).

M-R-S A-92 Diesel No Test
1974
Advertised @ 238 hp @ 2100 rpm, 4 1/2" x 5";
426 cu in. 6 cyl GM engine;
10 forward gears, 2.4 to 19.6 mph;
28,000 lbs.

Steiger Cougar II Diesel 1170
1974
227.31 db hp @ 2200 rpm, 4 3/4" x 6";
638 cu in. 6 cyl turbocharged caterpillar engine;
10 forward gears, 1.7 to 15 mph;
12.77 hp hr/gal; 26,690 lbs, 81.5 dB(A).

Steiger Panther II　　　　　　　　　　　　　　**No Test**
1974-
Manufacturer advertised 310 hp @ 2100 rpm, 5 1/2'' x 6'';
855 cu in. 6 cyl turbocharged Cummins engine;
10 forward gears, 2.2 to 19.1 mph;
26,000 lbs.

Steiger Wildcat II　　　　　　　　　　　　　　**No Test**
1974-
Advertised @ 185 hp @ 3000 rpm, 4.5'' x 5'';
636 cu in. V8 cyl Caterpillar engine;
10 forward gears, 2 to 16 mph;
18,000 lbs.

Steiger Tiger II Diesel　　　　　　　　　　　　1169
1974-
270.21 db hp @ 2600 rpm, 5.5'' x 4.75'';
903 cu in. V8 cyl turbocharged Cummins engine;
10 forward gears, 2 to 17.6 mph;
12.70 hp hr/gal; 27,600 lbs, 86.0 dB(A).

White Field Boss 4-150 Diesel　　　　　　　　1159
1974
151.87-131.13 hp @ 2800 rpm, 4 1/2'' x 5'';
636 cu in. V8 cyl Caterpiller engine;
18 forward gears, 1.6 to 19.2 mph;
11.34 hp hr/gal; 16,960 lbs, 81.0 dB(A).

FARM TRACTORS 1950-1975

1975

White Sets up Test Center

The White Farm Equipment Company established a research and test center at Libertyville, Illinois in June, 1975 and consolidated its engineering operations there.

About the same time, White introduced several new models, including the White 4-180. This 4-wheel drive tractor used a Caterpillar 636 cu in. diesel engine developing 181 PTO horsepower.

New 2-wheel drive White tractors included the 2-50, the 2-60, and the 2-85. Both the 2-50 and the 2-60 were made by Fiat with final assembly in Atlanta where tires, sheet metal, and electrical systems were added. The 2-50 had a 3-cylinder diesel engine which developed 47 horsepower; the 2-60 a 4-cylinder diesel which developed

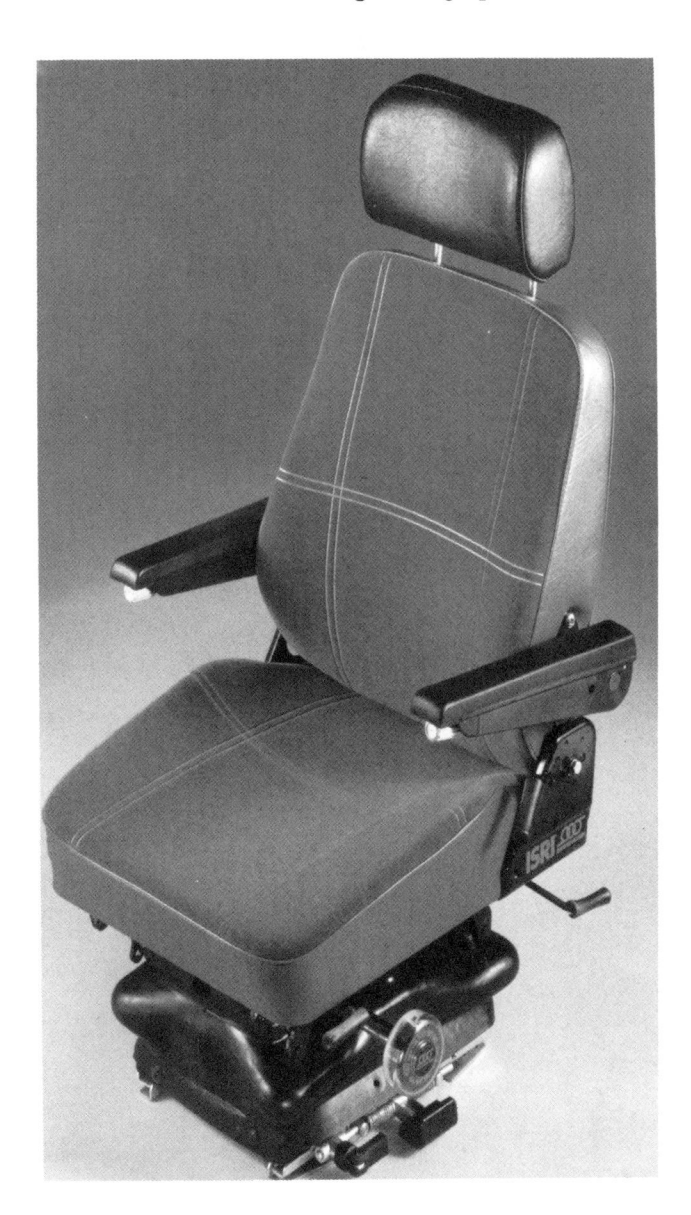

Similar seats are used on many modern tractors. This one is manufactured by Isringhausen, Battle Creek, Michigan.

Slope Tractor, Inc., Harper, Kansas manufactured a leaning wheel tractor that would operate on hillsides. The axles could tilt left or right 30 degrees, while the rest of the tractor remained vertical. The Slope Runner 1 weighed 5400 lbs and was equipped with a 4-cylinder Ford industrial engine, either gasoline or diesel. The transmission contained an 8-speed gear box giving a range of travel speeds from 1 to 18 mph. Steering was by power assist.

1950-1975 FARM TRACTORS

165

PERFORMANCE OF WHEEL TYPE TRACTORS
(10 HR. RUN)

DRAWBAR HP-HR PER GAL.

DIESEL

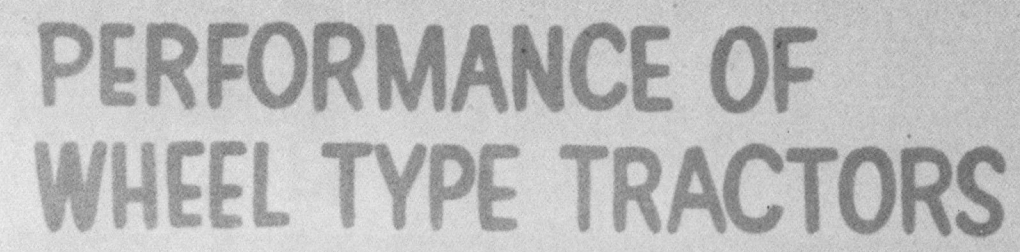

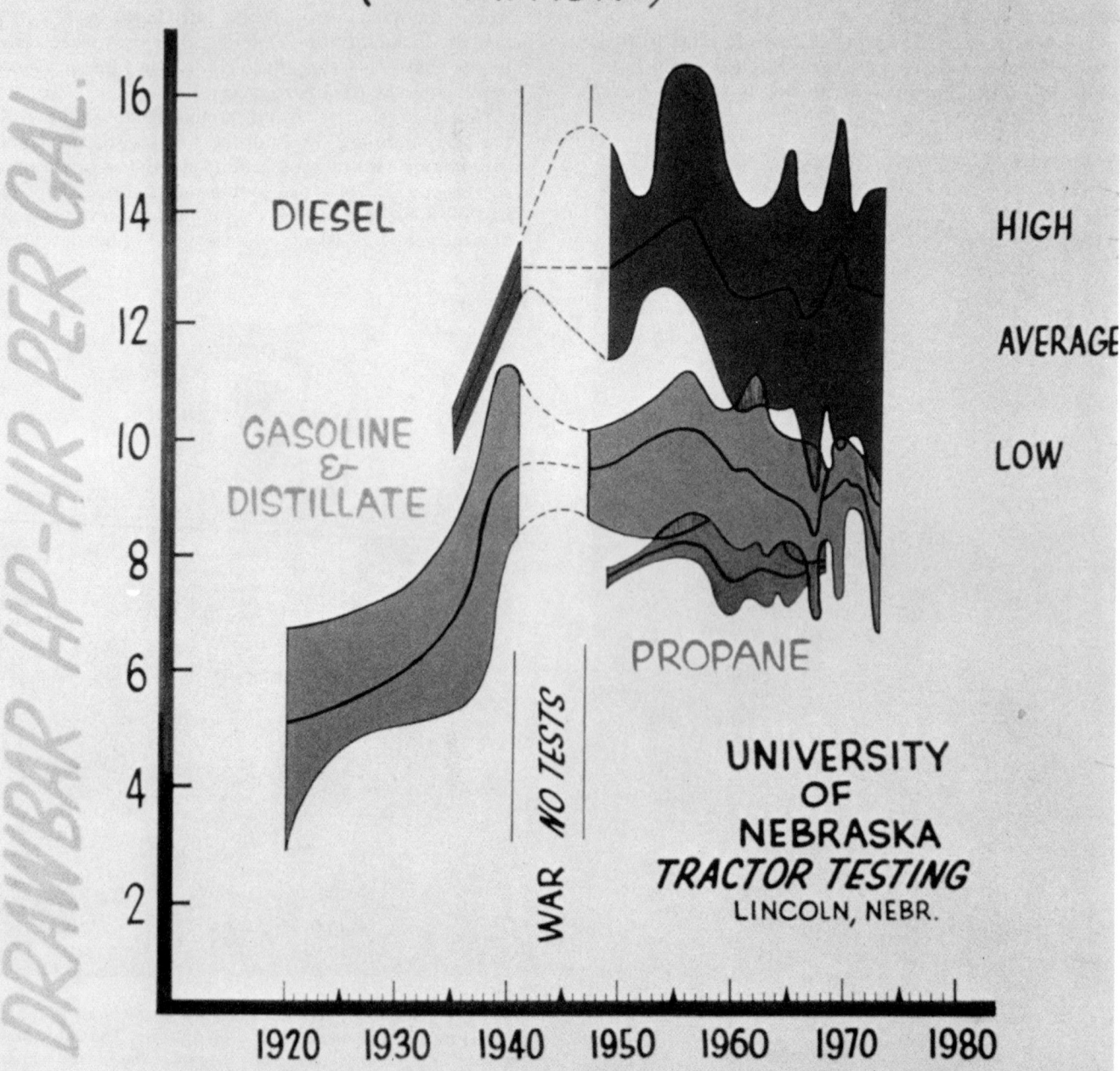

GASOLINE & DISTILLATE

PROPANE

WAR NO TESTS

HIGH
AVERAGE
LOW

UNIVERSITY OF NEBRASKA
TRACTOR TESTING
LINCOLN, NEBR.

DEPT. OF AGRICULTURAL ENGINEERING

FARM TRACTORS 1950-1975

The Slope Runner 2 weighed 2620 lbs and also was equipped with a Ford 4-cylinder industrial gasoline engine. The transmission was the hydrostatic type with a hydraulic motor in each rear wheel. Travel speed ranged up to 7 mph.

63 horsepower. The White 2-85 used a 354 cu in. diesel with 85 PTO horsepower.

Allis-Chalmers re-entered the smaller tractor competition with its 5040 Diesel, also Fiat-built. The tractor developed about 40 horsepower. In the larger classes, the firm replaced its Model 200 Diesel with the Model 7000 Diesel, which developed about 13 hp more than its predecessor.

Two other Allis-Chalmers diesel tractors, Models 7040 and 7060, were equipped with a power shift transmission providing six forward speeds and on-the-go power shifting through two mechanical gear ranges. Engines of both tractors were equipped with turbochargers and the 7060 also had an innercooler, which boosted its maximum horsepower by about 25 over the Model 7040.

Assembled by Steiger

International Harvester's second entry into the articulated 4-wheel drive field was the Model 4568, assembled by Steiger. This tractor carried an International turbocharged V-8 diesel engine with 800 cu in. displacement and 300 hp at 2600 rpm.

A new Massey-Ferguson 235 Diesel replaced the Model 135. A companion model with a gasoline engine had about the same horsepower.

John Deere introduced four new utility tractors, with

International Harvester Series 86 cab interior seat is well ahead of rear axle.

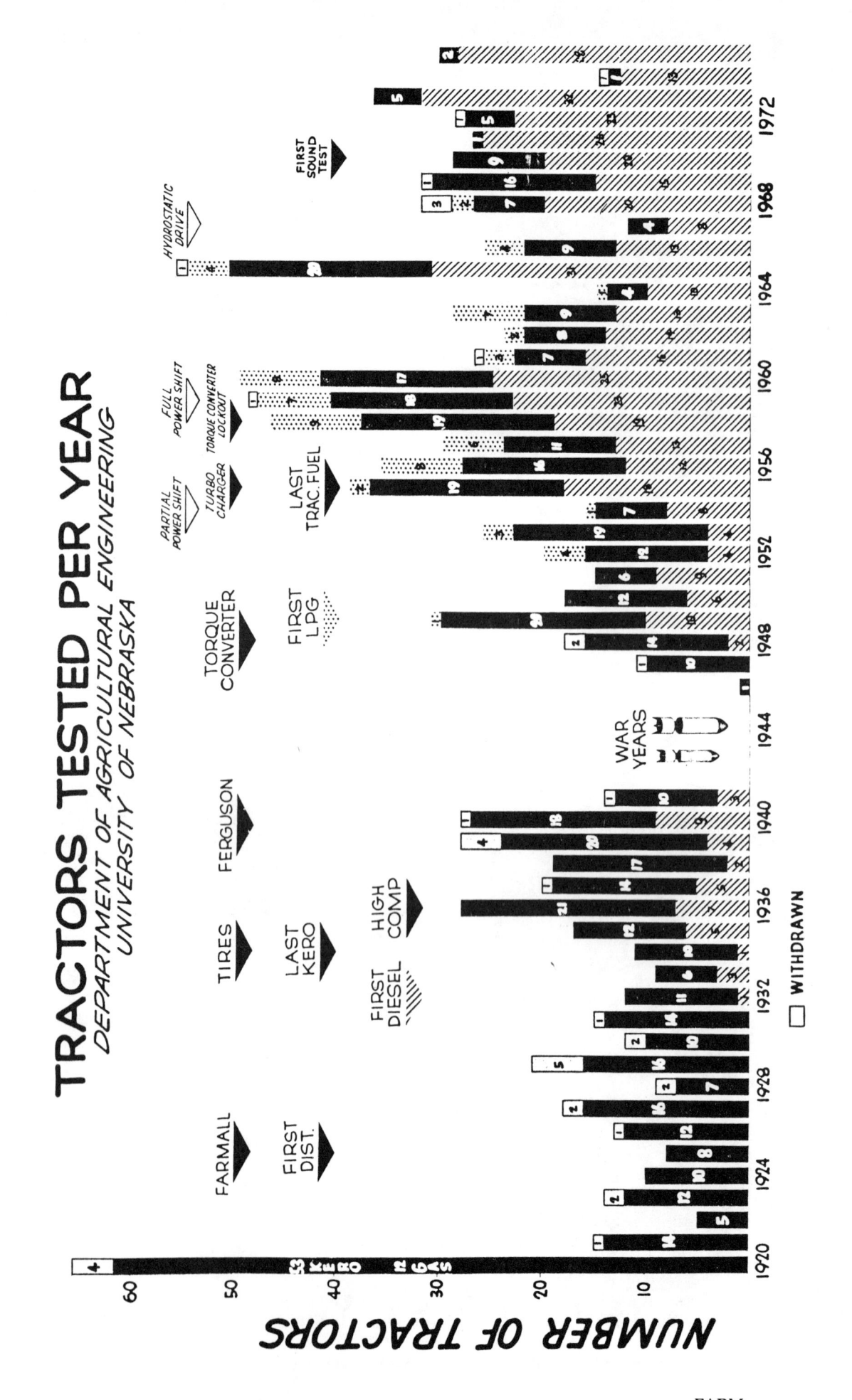

TRACTORS TESTED PER YEAR
DEPARTMENT OF AGRICULTURAL ENGINEERING
UNIVERSITY OF NEBRASKA

FARM TRACTORS 1950-1975

horsepower ranging from 40 to 70. They were the Models 2040, 2240, 2440, and 2640.

In compliance with British requirements, Leyland Tractors equipped all its vehicles with safety cabs. The new cabs also reduced the noise level for the operator.

Kubota, Ltd., of Japan, set up liaison offices in Carson City, California in 1975. The same year it introduced three new tractors—B-6000, the L225, and the L285.

The Kubota B-6000 was available in either 2- or 4-wheel drive. It was powered by a 12.5 hp diesel. The L225 was powered by a 24 hp diesel and the L285 by a 30 hp diesel.

New Case model tractors announced for 1975 included the Model 1410 and 1412 tractors built by David Brown. The Unimog Diesel manufactured by Daimler-Benz A.G., Stuttgart, West Germany was added to the Case line as the Case M-B 4/94 (Test 1211) in North America. A smaller version of the Unimog had been tested at Nebraska in 1957 (Test 607). Both vehicles were 4-wheel drive and could serve either as a tractor or truck. The M-B 4/94 was available with transmission options providing up to 20 forward speeds.

The new Massey-Ferguson 235 Diesel replaced Model 135. A companion, Model 235 gasoline, was available with nearly the same horsepower.

A new Model 255 Diesel, the 265 Diesel, the 275 (replacing the 175) and the 285 Diesel were also introduced.

Two other large 4-wheel drive tractors were also announced. These were the 1505 and the 1805. The 1805 replaced the earlier 1800 with increased power. Both tractors had fully independent PTO.

LOOKING BACK

These past 25 years were significant for the changes they brought in the tractor industry. Fuels changed. Economy records were made. Gasoline declined in usage, distillate disappeared, LPG appeared for a time, while diesel proved to be the predominant fuel.

Demands by farmers seeking more power for larger acreages and more aggressive agriculture caused tractor manufacturers to increase engine displacement, use higher engine speeds, install turbochargers and other methods to dramatically increase horsepower.

Many of the improvements were gains which we now take for granted. Four-wheel drive tractors were revived and have gained in popularity. New types of transmissions, operator enclosures, new safety features, noise reduction, power steering, better instrumentation, operator comfort and convenience plus many other features; all were conceived during the past 25-year period.

One can only wonder what the author of *FARM TRACTORS: 1975-2000* will have to report.

Allis-Chalmers 5040 Diesel   1230
1975
40.05-35.86 hp @ 2400 rpm, 3.74'' x 4.33'';
143 cu in. 3 cyl Uzina engine;
9 forward gears, 0.5 to 15.2 mph;
12.74 hp hr/gal; 4060 lbs, 95.5 dB(A).

Manufactured by Uzina Tractorul Brasov (UTB), Romania.

Allis-Chalmers 7580 Diesel   1229
1975
186.35-160.58 hp @ 2550 rpm, 4 1/4'' x 5.0'';
426 cu in. 6 cyl turbocharged Allis-Chalmers engine with intercooler;
20 forward gears, 1.5 to 17.6 mph;
11.39 hp hr/gal; 23,510 lbs, 81.0 dB(A).

1950-1975   FARM TRACTORS

Belarus 250 Diesel — 1183
1975
24.95-21.81 hp @ 1800 rpm, 4.134" x 4.724";
127 cu in. 2 cyl air-cooled Vladimir engine;
12 forward gears, 1.1 to 13.6 mph;
13.20 hp hr/gal; 4551 lbs, 85.0 dB(A).

Built in USSR at Vladimir Tractor Plant.

Belarus 7010 4-WD Diesel — No Test
1975
Advertised 235 hp @ 1700 rpm, 5.12" x 5.51";
907 cu in. V8 cyl Yamz turbocharged engine;
16 forward gears, 1.6 to 18.7 mph;
26,015 lbs.

Selective gear fixed ratio partial range operator-controlled power-shifting.

Ample seating space for two people in cab.

Manufactured at Leningrad Tractor plant, USSR.

Big Bud KT 450 — No Test
1975
Manufacturer advertised 450 hp @ 2100 rpm, 6.0" x 6.0";
1150 cu in. 6 cyl turbocharged Cummins engine;
13 forward gears, 1.3 to 16.6 mph;
43,000 lbs, advertised 75 dB(A).

J.I. Case M-B 4/94 Diesel — 1211
1975
73.33-66.33 hp @ 2550 rpm, 3.82" x 5.04";
346 cu in. 6 cyl Unimog engine;
12 forward gears, 2.4 to 46 mph;
11.65 hp hr/gal; 8490 lbs, 86.0 dB(A).

Manufactured by Mercedes-Benz and marketed in North America by Case. Primary use is in construction market.

FARM TRACTORS 1950-1975

John Deere 2040 Diesel   1191
1975
40.86-34.15 hp @ 2500 rpm, 4.02" x 4.33";
164 cu in. 3 cyl John Deere engine;
8 forward gears, 1.5 to 17.6 mph;
11.15 hp hr/gal; 4800 lbs, 96.5 dB(A).

Manufactured at John Deere Werke, Mannheim, West Germany.

John Deere 2240 Diesel   1192
1975
50.37-41.28 hp @ 2500 rpm, 4.19" x 4.33";
179 cu in. 3 cyl John Deere engine;
16 forward gears, 1.2 to 17.6 mph;
11.18 hp hr/gal; 4950 lbs, 98.0 dB(A).

Manufactured at John Deere Werke, Mannheim, West Germany.

John Deere 2440 Diesel   **No Test**
1975
Manufacturer's estimate of 60 hp @ 2500 rpm, 4.02" x 4.33";
219 cu in. 4 cyl John Deere engine;
16 forward gears, 1.1 to 17.8 mph.

Same as Model 2030 Diesel (Test 1085), except for styling.

John Deere 2640 Diesel   1157
1974
70.37-58.15 hp @ 2500 rpm, 4.19" x 5.0";
276 cu in. 4 cyl John Deere engine;
16 forward gears, 1.3 to 16.3 mph;
10.98 hp hr/gal; 5800 lbs, 97.0 dB(A).

1950-1975 FARM TRACTORS

Deutz Intrac 2003 No Test
1975
2300 rpm, 3.94" x 4.72";
230 cu in. 4 cyl air-cooled Deutz engine;
12 forward gears, 0.2 to 15.5 mph;
6500 lbs.

Four-wheel drive with three-point hitch at front and rear.

Seating space for two people in cab. PTO at front and rear. Engine mounted mid-ship.

Ford 9600 Diesel 1122
1975
135.46-114.34 hp @ 2200 rpm, 4.4" x 4.4";
401 cu in. 6 cyl turbocharged Ford engine;
16 forward gears, 1.5 to 17.6 mph;
12.64 hp hr/gal; 11,860 lbs, 85.0 dB(A).

International 4568 Diesel 1216
1975-1976
228.01-235.72 hp @ 2600 rpm, 5 5/16" x 4.5";
798 cu in. V8 cyl turbocharged International Harvester engine;
10 forward gears, 2.6 to 22.8 mph;
11.79 hp hr/gal; 25,750 lbs, 81.0 dB(A).

Kubota L285 Diesel 1198
1975-
26.45-22.94 hp @ 2400 rpm, 2.992" x 3.228";
91 cu in. 4 cyl Kubota engine;
8 forward gears, 1.2 to 12.3 mph;
12.16 hp hr/gal; 2445 lbs, 89.5 dB(A).

FARM TRACTORS 1950-1975

Leyland 245 Diesel   1177
1975-
39.73-35.09 hp @ 2250 rpm, 3.6" x 5.0";
152.7 cu in. 3 cyl Leyland engine;
10 forward gears, 1.3 to 16.2 mph;
12.67 hp hr/gal; 5880 lbs, 99.5 dB(A).

Manufactured by British Leyland Ltd., Bathgate, West Lothian, Scotland.

Leyland 255 Diesel   1176
1975
47.91-42.86 hp @ 2200 rpm, 3.82" x 4.83";
222 cu in. 4 cyl Leyland engine;
10 forward gears, 1.4 to 17.6 mph;
11.81 hp hr/gal; 6300 lbs, 97.5 dB(A).

Leyland 2100 Diesel   1178
1975
88.40-75.70 hp @ 2100 rpm, 3.86" x 4.92";
344.7 cu in. 6 cyl Leyland engine;
10 forward gears, 1.6 to 19.3 mph;
13.24 hp hr/gal; 8910 lbs, 91.5 dB(A).

Massey-Ferguson 235 Diesel   1187
1975-1976
42.39-35.62 hp @ 2250 rpm, 3.6" x 5.0";
153 cu in. 3 cyl Perkins engine;
8 forward gears, 1.6 to 20.8 mph;
13.39 hp hr/gal; 4135 lbs, 98.5 dB(A).

Model 235 Gasoline (Test 1194) with 145 cu in. Continental engine and same rpm developed 41.13 hp and 8.07 hp hr/gal. 94.5 dB(A)

1950-1975 FARM TRACTORS

Massey-Ferguson 255 Diesel   1189
1975-1977
50.69-43.72 hp @ 2000 rpm, 3.6" x 5.0";
203 cu in. 4 cyl Perkins engine;
12 forward gears, 1.4 to 19 mph;
13.36 hp hr/gal; 5970 lbs, 95.5 dB(A).

Model 255 Gasoline (Test 1190) with 212 cu in. displacement and same rpm developed 50.01 hp and 8.54 hp hr/gal. 94.0 dB(A)

Massey-Ferguson 265 Diesel   1188
1975-1977
60.73-51.87 hp @ 2000 rpm, 3.875" x 5.0";
236 cu in. 4 cyl Perkins engine;
12 forward gears, 1.4 to 19 mph;
13.51 hp hr/gal; 6385 lbs, 96.0 dB(A).

Massey-Ferguson 275 Diesel   1193
1975-1977
67.43-60.16 hp @ 2000 rpm, 3.975" x 5.0";
248 cu in. 4 cyl Perkins engine;
12 forward gears, 1.3 to 19.2 mph;
13.22 hp hr/gal; 6610 lbs, 98.5 dB(A).

Massey-Ferguson 285 Diesel   1171
1975-1977
81.96-70.07 hp @ 2000 rpm, 4.5" x 5.0";
318 cu in. 4 cyl Perkins engine;
12 forward gears, 1.4 to 19.6 mph;
12.90 hp hr/gal; 7650 lbs, 97.5 dB(A).

Massey-Ferguson 1505 Diesel   1172
1975-1977
174.46-148.22 hp @ 2800 rpm, 4.5" x 5.0";
636 cu in. V8 cyl Caterpillar engine;
12 forward gears, 2.1 to 19.8 mph;
11.43 hp hr/gal; 17,960 lbs, 85.5 dB(A).

Massey-Ferguson 1805 Diesel   1173
1975-1977
192.65-167.06 hp @ 2800 rpm, 4.5" x 5.0";
636 cu in. V8 cyl Caterpillar engine;
12 forward gears, 2.1 to 19.8 mph;
12.41 hp hr/gal; 18,000 lbs, 86.5 dB(A).

SAME Buffalo 4-WD Diesel   1186
1975
104.94-87.24 hp @ 2200 rpm, 4.13" x 4.72";
380 cu in. 6 cyl SAME air-cooled engine;
12 forward gears, 1.1 to 13.7 mph;
12.44 hp hr/gal; 10,087 lbs, 97.5 dB(A).

The Model SAME Panter 4WD Diesel (Test 1185) with a 5-cylinder (276 cu in.) engine developed 80.02 max hp and 12.30 hp hr/gal.

SAME Panter 4-WD Diesel   1185
1975
80.02-66.67 hp @ 2200 rpm, 3.86" x 4.72";
276 cu in. 5 cyl SAME air-cooled engine;
12 forward gears, 0.5 to 14.3 mph;
12.30 hp hr/gal; 9228 lbs, 96.0 dB(A).

Manufactured by SAME S.p.A., Treviglio, Italy.

Selective gear fixed ratio with operator-control partial range power-shift.

Versatile (Series 2) 800 Diesel　　　　No Test
1975
Advertised @ 235-179 hp @ 2100 rpm, 5.5'' x 6.0'';
855 cu in. 6 cyl Cummins engine;
12 forward gears, 2.6 to 14.3 mph;
19,500 lbs.

White 2-150 Diesel　　　　1182
1975
147.49-124.92 hp @ 2200 rpm, 4 3/4'' x 5.5'';
585 cu in. 6 cyl White engine;
18 forward gears, 1.5 to 17.7 mph;
12.14 hp hr/gal; 16,940 lbs, 90.0 dB(A).

White 2-105 Diesel　　　　1181
1975
105.61-87.23 hp @ 2200 rpm, 3 7/8'' x 5.0'';
354 cu in. 6 cyl turbocharged Perkins engine;
18 forward gears, 1.6 to 18.9 mph;
11.95 hp hr/gal; 11,810 lbs, 88.5 dB(A).

White Field Boss 4-180 Diesel　　　　1184
1975
181.07-156.54 hp @ 2801 rpm, 4 1/2'' x 5.0'';
636 cu in. V8 cyl Caterpillar engine;
12 forward gears, 1.9 to 18.9 mph;
11.56 hp hr/gal; 20,600 lbs, 83.5 dB(A).

Articulated steering 4-WD.

Yanmar YM 240 Diesel 1199
1975
19.76-16.60 hp @ 2400 rpm, 3.543'' x 3.543'';
70 cu in. 2 cyl Yanmar engine;
8 forward gears, 0.5 to 8.6 mph;
11.95 hp hr/gal; 1950 lbs, 88.5 dB(A).

Manufactured by Yanmar Diesel Engine Co. Ltd., Osaka, Japan.

APPENDIX
TRACTORS TESTED AT NEBRASKA
1950-1975

1950

- 437 Minneapolis-Moline G Gaso
- 438 Minneapolis-Moline Z Gaso
- 439 Allis-Chalmers B Gaso
- 440 Allis-Chalmers WD Gaso
- 441 McCormick-Deering WD-9 Dsl
- 442 Cockshutt 40 Gaso
- 443 Ford 8N Gaso
- 444 Ford 8NAN Tractor Fuel
- 445 International TD-14A Dsl
- 446 International TD-18A Dsl
- 447 International TD-24 Dsl
- 448 John Deere MC Gaso
- 449 Choremaster "B" Gaso
- 450 Oliver Row Crop 88 Dsl
- 451 Oliver 99 Gaso
- 452 Massey-Harris 55 Dsl
- 453 Allis-Chalmers CA Gaso
- 454 Dodge Power Wagon T137 Gaso

1951

- 455 Massey-Harris 55 Gaso
- 456 David Bradley Garden Tractor Gaso
- 457 Oliver Row Crop 77 Dsl
- 458 McCormick Farmall Super C Gaso
- 459 McCormick WD-6 Dsl
- 460 McCormick Farmall MD Dsl
- 461 International TD-9 Dsl
- 462 International TD-6 Dsl
- 463 Allis-Chalmers HD-9 Dsl
- 464 Allis-Chalmers HD-15 Dsl
- 465 Allis-Chalmers HD-20 Dsl
- 466 Ferguson TO-30 Gaso
- 467 Oliver Row Crop 66 Dsl
- 468 Minneapolis-Moline R Gaso
- 469 Minneapolis-Moline BF Gaso

1952

- 470 Oliver 77 LPG
- 471 Terratrac GT-30 Gaso
- 472 John Deere Model 60 Gaso
- 473 Bolens 12BB Gaso
- 474 Cockshutt 20 Gaso
- 475 McCormick Farmall Super M Gaso
- 476 McCormick Super W-6 Gaso
- 477 McCormick Farmall Super MD Dsl
- 478 McCormick Super WD-6 Dsl
- 479 Harris Power Horse 53 Gaso
- 480 Case LA Gaso
- 481 Case LA Tractor Fuel
- 482 Case LA LPG
- 483 Economy Special Gaso
- 484 McCormick Farmall Super M LPG
- 485 McCormick Super W-6 LPG
- 486 John Deere Model 50 Gaso
- 487 Cockshutt 50 Dsl
- 488 Cockshutt 50 Gaso
- 489 Oliver OC-18 Dsl

1953

- 490 John Deere 60 All Fuel Tractor Fl
- 491 McCormick Super W-4 Gaso
- 492 Farmall Super H Gaso
- 493 John Deere 70 Gaso
- 494 Ford NAA Gaso
- 495 Someca DA 50 Dsl
- 496 Case SC Gaso
- 497 Case SC Tractor Fuel
- 498 Intercontinental DF Dsl
- 499 Allis-Chalmers WD 45 Gaso
- 500 New Fordson Major Dsl
- 501 New Fordson Major Gaso
- 502 Willys Farm Jeep Gaso
- 503 John Deere 40 Gaso
- 504 John Deere 40 S Gaso
- 505 John Deere 40 C Gaso
- 506 John Deere 70 All Fuel Tractor Fl
- 507 John Deere 50 All Fuel Tractor Fl
- 508 Case Model 500 Dsl
- 509 Massey-Harris 33 RT Gaso
- 510 Massey-Harris 44 Special Gaso
- 511 Allis-Chalmers WD-45 Dist
- 512 Allis-Chalmers WD-45 LPG
- 513 John Deere 60 LPG
- 514 John Deere 70 LPG
- 515 David Bradley Super Power Gaso

1954

- 516 Oliver OC-5 Gaso
- 517 Oliver OC-6 Dsl
- 518 McCormick Super WD-9 Dsl
- 519 Harris Four-Wheel Drive FDW-C Dsl
- 520 Minneapolis-Moline UB Gaso
- 521 Minneapolis-Moline U Trac Fuel
- 522 Minneapolis-Moline UB LPG
- 523 Harris Four-Wheel Drive FDW-C Dsl
- 524 Oliver Super 55 HC Gaso
- 525 Oliver Super 88 HC Gaso
- 526 Oliver Super 55 Dsl
- 527 Oliver Super 88 Dsl
- 528 John Deere 70 Dsl
- 529 International TD-24 Dsl
- 530 David Brown 25 Gaso
- 531 Massey-Harris No. 16 Pacer Gaso

1955

- 532 McCormick Farmall 400 Gaso
- 533 International W-400 Gaso
- 534 McCormick Farmall 400 Dsl
- 535 International W-400 Dsl
- 536 McCormick Farmall 200 Gaso
- 537 McCormick Farmall 100 Gaso
- 538 McCormick Farmall 300 Gaso
- 539 International 300 Utility Gaso
- 540 John Deere 50 LPG
- 541 Oliver Super 66 HC Gaso
- 542 Oliver Super 77 HC Gaso
- 543 Oliver Super 77 Dsl
- 544 Oliver Super 66 Dsl
- 545 Minneapolis-Moline GB LPG
- 546 John Deere 40-S All Fuel Tractor Fl
- 547 Minneapolis-Moline GB Gaso
- 548 Oliver OC-12 Gaso
- 549 Oliver OC-12 Dsl
- 550 Allis-Chalmers HD-21 Dsl

178

FARM TRACTORS 1950-1975

551	Allis-Chalmers HD-16 Dsl		614	Case 301 Dsl
552	Allis-Chalmers HD-16 A Dsl		615	International 350 Utility Gaso
553	Caterpillar D-2 Dsl		616	McCormick Farmall 230 Gaso
554	Caterpillar D-4 Dsl		617	McCormick Farmall 130 Gaso
555	Caterpillar D-6 Dsl		618	International 650 Gaso
556	Oliver Super 99 GM Dsl		619	International 350 Utility LPG
557	Oliver Super 99 Dsl		620	McCormick Farmall 450 LPG
558	Nuffield Universal DM-4 Dsl		621	International 650 LPG
559	Nuffield Universal PM-4 Gaso		622	McCormick Farmall 350 LPG
560	Ford 640 Gaso		623	Allis-Chalmers D-14 Gaso
561	Ford 660 Gaso		624	Minneapolis-Moline 335 Gaso
562	Ford 860 Gaso		625	Someca Som 45 Dsl
563	Allis-Chalmers WD-45 Dsl		626	Ford 850-L LPG
564	Ferguson TO-35 Gaso		627	Ford 640-L LPG
565	Case 401 Dsl		628	Eimco 105 Dsl
566	Case 411 Gaso		629	International TD-18 Dsl
567	John Deere 80 Dsl		630	International TD-24 Dsl
568	Minneapolis-Moline GB Dsl		631	Wagner TR-9 Dsl
569	Ford 960 Gaso		632	John Deere 820 Dsl
570	Ford 740 Gaso		633	Oliver OC-15 Dsl
			634	International 330 Utility Gaso
			635	Allis-Chalmers D-17 Gaso
			636	Allis-Chalmers D-17 Dsl

1956

571	McCormick Farmall 400 LPG
572	International W-400 LPG
573	McCormick Farmall 300 LPG
574	International 300 Utility LPG
575	McCormick Farmall CUB Gaso
576	Massey-Harris 444 Dsl
577	Massey-Harris 333 Dsl
578	Minneapolis-Moline 445 Univ Gaso
579	Minneapolis-Moline 445 Utility Gaso
580	Allis-Chalmers HD-6B Dsl
581	Allis-Chalmers HD-11B Dsl
582	Caterpillar D-7 Dsl
583	Caterpillar D-8 Dsl
584	Caterpillar D-9 Dsl
585	International TD-14 Dsl
586	International TD-9 Dsl
587	International TD-6 Dsl
588	International TD-18 Dsl
589	International T-6 Gaso
590	John Deere 520 LPG
591	John Deere 620 LPG
592	John Deere 520 All Fuel Tractor Fl
593	John Deere 720 LPG
594	John Deere 720 Dsl
595	Massey-Harris 50 Gaso
596	Ferguson 40 Gaso
597	John Deere 520 Gaso
598	John Deere 620 Gaso
599	John Deere 420-W Gaso
600	John Deere 420S All Fuel Tractor Fl
601	John Deere 420 "C" Gaso
602	Massey-Harris 444 LPG
603	Massey-Harris 333 Gaso
604	John Deere 620 All Fuel Tractor Fl
605	John Deere 720 Gaso
606	John Deere 720 All Fuel Tractor Fl

1957

607	Unimog 30 Dsl
608	McCormick Farmall 450 Dsl
609	McCormick Farmall 350 Dsl
610	International 350 Utility Dsl
611	McCormick Farmall 350 Gaso
612	McCormick Farmall 450 Gaso
613	Case 311 Gaso

1958

637	Volvo T425 Gaso
638	Volvo T55 Dsl
639	Ford 841 L LPG
640	Ford 851 Gaso
641	Ford 841 Gaso
642	Ford 641 Gaso
643	Ford 651 Gaso
644	Allis-Chalmers D-17 LPG
645	Allis-Chalmers D-14 LPG
646	Case 511B Gaso
647	Oliver 880 Gaso
648	Oliver 770 Gaso
649	Oliver 770 Dsl
650	Oliver 880 Dsl
651	Minneapolis-Moline 5 Star LPG
652	Minneapolis-Moline 5 Star Dsl
653	Ford 841 Dsl
654	Ford 851 Dsl
655	Oliver OC-4 Dsl
656	Oliver OC-4 Gaso
657	Massey-Ferguson MF-65 LPG
658	Massey-Ferguson MF-50 LPG
659	Massey-Ferguson MF-65 Gaso
660	Oliver 950 Dsl
661	Oliver 990 GM Dsl
662	Oliver 995 GM Lugmatic Dsl
663	Nuffield Universal Three Dsl
664	Allis-Chalmers HD-21A Dsl
665	McCormick Farmall 340 Gaso
666	McCormick Farmall 140 Gaso
667	McCormick Farmall 240 Gaso
668	International 240 Utility Gaso
669	McCormick Farmall 560 Dsl
670	McCormick Farmall 460 Gaso
671	McCormick Farmall 560 Gaso
672	McCormick Farmall 460 Dsl
673	International 460 Utility Dsl
674	International 460 Utility Gaso
675	McCormick Farmall 560 LPG
676	McCormick Farmall 460 LPG
677	International 460 Utility LPG
678	Case 711-B Gaso

679	Case 811-B Gaso
680	Case 801-B Dsl
681	Cockshutt 550 Dsl
682	Cockshutt 560 Dsl
683	Cockshutt 570 Dsl

1959

684	Fordson Dexta Dsl
685	Fordson Power Major Dsl
686	Ford 641-D Dsl
687	Case 611-B Gaso
688	Case 211-B Gaso
689	Case 411-B Gaso
690	Massey-Ferguson TO-35 Dsl
691	David Brown 950 Dsl
692	Case 900-B Dsl
693	Case 701-B Dsl
694	Case 711-B LPG
695	Case 811-B LPG
696	Case 910-B LPG
697	Oliver 550 Gaso
698	Oliver 550 Dsl
699	Porsche Junior L108 Dsl
700	Wagner TR-14A Dsl
701	Ford 881 Gaso
702	Ford 681 Gaso
703	Ford 881 LPG
704	Ford 681 LPG
705	Ford 881 Dsl
706	Ford 681 Dsl
707	Fiat 411-R Dsl
708	Fiat 411-C Dsl
709	Case 310-C Gaso
710	Caterpillar D-7 Dsl
711	Caterpillar D-8 Dsl
712	Case 1010 Dsl
713	Case 810 Dsl
714	Case 610 Dsl
715	International 660 Dsl
716	John Deere 435 Dsl
717	John Deere 440 ID Dsl
718	John Deere 440 I Gaso
719	John Deere 440 ICD Dsl
720	John Deere 440 IC Gaso
721	International 660 Gaso
722	International 660 LPG
723	Allis-Chalmers D-12 Gaso
724	Allis-Chalmers D-10 Gaso
725	International T-340 Gaso
726	Massey-Ferguson 85 Gaso
727	Massey-Ferguson 85 LPG
728	Porsche Super L 318 Dsl
729	David Bradley Suburban Gaso
730	David Bradley Super 575 Gaso
731	David Bradley Super 300 Gaso
732	Bradley Handiman Gaso

1960

733	McCormick International B-275 Dsl
734	David Brown 850 Dsl
735	David Brown 850 Gaso
736	Case 831C Dsl
737	Case 831 Dsl
738	Case 841 Gaso
739	Case 940 LPG
740	Case 841 LPG

741	Case 930 Dsl
742	Case 731C Dsl
743	Case 731 Dsl
744	Massey-Ferguson MF 35 Dsl
745	Massey-Ferguson MF 65 Dsl
746	Caterpillar D4 Dsl
747	Caterpillar D6 Dsl
748	Zetor 50 Super Dsl
749	Land-Rover 88 Gaso
750	International TD-15 Dsl
751	International TD-9 Dsl
752	International TD-25 Dsl
753	International T-5 Gaso
754	International T-4 Gaso
755	International TD-5 Dsl
756	Minneapolis-Moline M-5 Gaso
757	Minneapolis-Moline M-5 LPG
758	Minneapolis-Moline M-5 Dsl
759	John Deere 4010 Gaso
760	John Deere 4010 LPG
761	John Deere 4010 Dsl
762	John Deere 3010 Dsl
763	John Deere 3010 Gaso
764	John Deere 3010 LPG
765	Massey-Ferguson 88 Dsl
766	Oliver 1800 Gaso
767	Oliver 1800 Dsl
768	Oliver 1900 Dsl
769	Case 541C Dsl
770	Case 640C Gaso
771	Case 541 Gaso
772	Case 531 Dsl
773	Case 640 Gaso
774	Case 441 Gaso
775	McCormick Farmall 340 Dsl
776	International TD 340 Dsl
777	Case 841C Gaso
778	Case 741C Gaso
779	Case 741 Gaso
780	Case 841C LPG
781	Case 741C LPG
782	Case 741 LPG

1961

783	Ford 6000 Dsl
784	Ford 6000 Gaso
785	Case 431 Dsl
786	Case 531C Dsl
787	Case 630C Dsl
788	Case 630 Dsl
789	Minneapolis-Moline 4 Star Gaso
790	Minneapolis-Moline 4 Star LPG
791	Minneapolis-Moline Gvi LPG
792	Minneapolis-Moline Gvi Dsl
793	Allis-Chalmers H3 Gaso
794	Allis-Chalmers HD3 Dsl
795	Allis-Chalmers D15 Gaso
796	Allis-Chalmers D15 Dsl
797	Allis-Chalmers D15 LPG
798	John Deere 1010 C Dsl
799	John Deere 2010 RU Dsl
800	John Deere 2010 RU Gaso
801	John Deere 1010C Gaso
802	John Deere 1010 RU Gaso
803	John Deere 1010 RU Dsl
804	Ursus C-325 Dsl
805	Case 310E Dsl

FARM TRACTORS 1950-1975

806 International TD-20 Dsl
807 Massey-Ferguson MF50 Dsl
808 Massey-Ferguson MF65 Dieselmatic

1962

809 Nuffield 460 Dsl
810 Allis-Chalmers D19 Gaso
811 Allis-Chalmers D19 Dsl
812 Allis-Chalmers D10 Gaso
813 Allis-Chalmers D12 Gaso
814 Allis-Chalmers D19 LPG
815 International 4300 Dsl
816 IHC Farmall 504 Dsl
817 International B414 Gaso
818 IHC Farmall 404 Gaso
819 IHC Farmall 504 Gaso
820 IHC Farmall 504 LPG
821 Kramer KL 400 Dsl
822 Massey-Ferguson Super 90 Dsl
823 Zetor 3011 Dsl
824 Oliver 1900 Series B Dsl
825 International 606 Gaso
826 International 606 Dsl
827 International B414 Dsl
828 John Deere 5010 Dsl
829 John Deere 2010C Gaso
830 John Deere 2010C Dsl
831 Oliver 1800 Series B Dsl
832 Oliver 1800 Series B 4-WD Dsl

1963

833 Minneapolis-Moline G706 Dsl
834 Minneapolis-Moline G706 LPG
835 Minneapolis-Moline G705 Dsl
836 Minneapolis-Moline G705 LPG
837 Allis-Chalmers Series II D15 Gaso
838 Allis-Chalmers Series II D15 LPG
839 Oliver 1800 Series B Gaso
840 Oliver 1600 Dsl
841 Oliver 1600 Gaso
842 Ford 6000 Gaso
843 Ford 6000 Dsl
844 Ford 2000 Super Dexta Dsl
845 Ford 5000 Dsl
846 Oliver 1800 Series B 4-WD Gaso
847 Oliver 1900 Series B 4-WD Dsl
848 John Deere 3020 Dsl
849 John Deere 4020 Dsl
850 John Deere 4020 Gaso
851 John Deere 3020 Gaso
852 John Deere 3020 LPG
853 John Deere 4020 LPG
854 Massey-Ferguson MF 25 Dsl
855 Allis-Chalmers D 21 Dsl
856 Farmall 706 Dsl
857 Farmall 806 Dsl
858 Farmall 706 Gaso
859 Farmall 806 Gaso
860 Farmall 706 LPG
861 Farmall 806 LPG

1964

862 Minneapolis-Moline U302 Gaso
863 Minneapolis-Moline U302 LPG
864 FWD Wagner WA4 Dsl

865 Case 750 Dsl
866 Zetor 2011 Dsl
867 Zetor 4011 Dsl
868 Case 1200 Dsl
869 Oliver 1850 4-WD Dsl
870 Oliver 1850 Dsl
871 Oliver 1950 Dsl
872 Oliver 1950 4-WD Dsl
873 Oliver 1650 Dsl
874 Oliver 1650 Gaso
875 Oliver 1850 Gaso
876 Oliver 1850 4-WD Gaso

1965

877 Ford Commander 6000 Gaso
878 Ford Commander 6000 Dsl
879 Ford 5000 8-Speed Dsl
880 Ford 5000 Select-O-Speed Dsl
881 Ford 3000 8-Speed Dsl
882 Ford 3000 Select-O-Speed Dsl
883 Ford 3000 4-Speed Dsl
884 Ford 2000 4-Speed Gaso
885 Ford 3000 Select-O-Speed Gaso
886 Allis-Chalmers One-Ninety Dsl
887 Allis-Chalmers One-Ninety XT Dsl
888 Ford 3000 8-Speed Gaso
889 Ford 3000 4-Speed Gaso
890 Ford 4000 Select-O-Speed Gaso
891 Ford 4000 8-Speed Gaso
892 Ford 4000 Select-O-Speed Dsl
893 Ford 4000 8-Speed Dsl
894 Ford 2000 8-Speed Gaso
895 Massey-Ferguson MF135 Dsl
896 Massey-Ferguson MF165 Dsl
897 Massey-Ferguson MF175 Dsl
898 Massey-Ferguson MF165 Gaso
899 Massey-Ferguson MF135 Gaso
900 Massey-Ferguson MF180 Dsl
901 Massey-Ferguson MF150 Dsl
902 David Brown 880 Dsl
903 David Brown 990 Dsl
904 Allis-Chalmers Series II D21 Dsl
905 Nuffield 10/42 Dsl
906 Kubota RV Gaso
907 Nuffield 10/60 Dsl
908 International 424 Gaso
909 Farmall 656 Gaso
910 Farmall 1206 Turbo Dsl
911 International 424 Dsl
912 Farmall 656 Dsl
913 John Deere 2510 Power-Shift Gaso
914 John Deere 2510 Syncro-Range Gaso
915 John Deere 2510 Power-Shift Dsl
916 John Deere 2510 Syncro-Range Dsl
917 Case 831 CK (Also 830 CK & 832 CK) Dsl
918 Case 931 GP Dsl
919 Case 841 CK Gaso
920 Case 941 GP Gaso
921 Case 841CK LPG (Also 840 CK & 842 CK)
922 Case 941 GP LPG
923 Massey-Ferguson MF 1100 Dsl
924 Minneapolis-Moline M670 LPG
    (Also M-M M670 Super)
925 Minneapolis-Moline M670 Gaso
    (Also M-M M670 Super)
926 Minneapolis-Moline M670 Dsl
    (Also M-M M670 Super)

1950-1975 FARM TRACTORS

927	Allis-Chalmers One-Ninety XT LPG	987	Oliver 2050 Row Crop Dsl
928	Allis-Chalmers One-Ninety Gaso	988	Withdrawn
929	Allis-Chalmers One-Ninety XT Gaso	989	Withdrawn
930	John Deere 4020 Syncro-Range Dsl	990	Withdrawn
931	International 4100 4-WD Dsl	991	John Deere 1520 Dsl
		992	John Deere 2520 Syncro-Range Dsl
		993	John Deere 2520 Power-Shift Dsl

1966

932	Ford 5000 8-Speed Gaso	994	Ford 4000 8-Speed Dsl
			(Also Ford 4000 8-Speed Row Crop)
933	Ford 5000 Select-O-Speed Gaso	995	Ford 4000 Select-O-Speed Dsl
934	John Deere 4020 Syncro-Range LPG		(Also 4000 Select-O-Speed Row Crop)
935	John Deere 1020 Gaso	996	Ford 5000 8-Speed Dsl
936	John Deere 2020 Gaso		(Also Ford 5000 8-Speed Row Crop)
937	John Deere 1020 Dsl	997	Ford 5000 Select-O-Speed Gaso
938	John Deere 2020 Dsl		(Also 5000 Select-O-Speed Row Crop)
939	John Deere 4020 Syncro-Range Gaso	998	Ford 5000 8-Speed Gaso
940	John Deere 3020 Syncro-Range Dsl		(Also Ford 5000 8-Speed Row Crop)
941	John Deere 3020 Syncro-Range Gaso	999	Ford 4000 Select-O-Speed Gaso
942	John Deere 3020 Syncro-Range LPG		(Also 4000 Select-O-Speed Row Crop)
943	Oliver 1550 Diesel	1000	Ford 5000 Select-O-Speed Dsl
944	Oliver 1550 Gaso		(Also 5000 Select-O-Speed Row Crop)
945	David Brown 770 Selectamatic Dsl	1001	Ford 4000 8-Speed Gaso
946	David Brown 880 Selectamatic Dsl		(Also 4000 8-Speed Row Crop)
947	John Deere 5020 Dsl		
948	Massey-Ferguson MF1130 Dsl		
949	Massey-Ferguson MF130 Dsl		

1969

950	International 500 Gaso	1002	John Deere 2520 Power-Shift Gaso
951	International 500 Dsl	1003	John Deere 2520 Syncro-Range Gaso
952	Case 1031 Dsl	1004	John Deere 1520 Gaso
953	Minneapolis-Moline G1000 Dsl	1005	Allis-Chalmers 180 Gaso
954	Minneapolis-Moline G1000 LPG	1006	Case 1470 Traction King Dsl
955	Farmall 706 Dsl	1007	International Farmall 544 Hydro Gaso
956	Farmall 706 LPG	1008	Massey-Ferguson MF165 Gaso
957	Farmall 706 Gaso	1009	Massey-Ferguson MF135 Gaso
		1010	John Deere 3020 Power-Shift Gaso

1967

958	Ford 2000 4-Speed Dsl	1011	John Deere 3020 Syncro-Range Gaso
959	Ford 2000 8-Speed Dsl	1012	John Deere 4020 Power-Shift Gaso
960	Zetor 5511 Zetormatic Dsl	1013	John Deere 4020 Syncro-Range Gaso
961	Oliver 1750 Gaso	1014	John Deere 4520 Power-Shift Dsl
962	Oliver 1750 Dsl	1015	John Deere 4520 Syncro-Range Dsl
963	Massey-Ferguson MF1100 Gaso	1016	Withdrawn
964	Allis-Chalmers 180 Dsl	1017	Alllis-Chalmers Two-Twenty Dsl
965	Allis-Chalmers 170 Dsl	1018	David Brown 3800 Selectamatic Gaso
966	Allis-Chalmers 170 Gaso	1019	David Brown 4600 Selectamatic Gaso
967	International Farmall 656 Hydro Dsl	1020	Massey-Ferguson MF175 Gaso
968	International Farmall 656 Hydro Gaso	1021	Massey-Ferguson MF180 Gaso
969	Oliver 1950-T Row Crop Dsl	1022	Massey-Ferguson MF1080 Dsl
		1023	John Deere 4000 Dsl
		1024	John Deere 4020 Power-Shift Dsl
		1025	John Deere 5020 Dsl

1968

970	International Farmall 856 Dsl	1026	Ford 8000 Dual Power Dsl
			(Also 8000 Dual Power Row Crop)
971	International Farmall 1256 Turbo Dsl	1027	Ford 9000 Dual Power Dsl
972	Oliver 1950-T 4-WD Dsl		(Also 9000 Dual Power Row Crop)
973	Ford 8000 Dsl	1028	Allis-Chalmers One-Sixty Dsl
974	Minneapolis-Moline G1000 Vista Dsl	1029	IHC Farmall 544 Hydro Dsl
975	Minneapolis-Moline G1000 Vista LPG		(Also IHC 544 Hydro & IHC 2544 Hydro)
976	David Brown Selectamatic 1200 Dsl	1030	Case 870 Power-Shift Dsl
977	David Brown Selectamatic 990 Dsl	1031	Case 870 Manual Dsl
978	Minneapolis-Moline G900 Dsl	1032	Case 770 Power-Shift Gaso
979	Minneapolis-Moline G900 LPG	1033	Case 770 Manual Dsl
980	Minneapolis-Moline G900 Gaso		
981	Ursus C-335 Dsl		

1970

982	Ursus C-350 Dsl	1034	Case 970 Manual Dsl
983	International Farmall 544 Dsl	1035	Case 1070 Manual Dsl
984	International Farmall 544 Gaso		(Also 1090 Manual)
985	International 444 Gaso	1036	Case 1070 Power-Shift Dsl
986	Oliver 2150 Row Crop Dsl		(Also 1090 Power-Shift)

FARM TRACTORS 1950-1975

1037	Case 970 Power-Shift Dsl
1038	Case 970 Power-Shift Gaso
1039	Case 970 Manual Gaso
1040	Oliver 1855 Dsl
1041	Oliver 1655 Dsl
1042	Oliver 1655 Gaso
1043	Allis-Chalmers Crop Hustler 175 Dsl
1044	Allis-Chalmers Crop Hustler 185 Dsl
1045	IHC Farmall 826 Dsl
	(Also IHC 826)
1046	IHC Farmall 826 Hydro Dsl
	(Also IHC 826 Hydro)
1047	IHC Farmall 1026 Hydro Dsl
	(Also IHC 1026 Hydro)
1048	IHC Farmall 1456 Dsl
	(Also IHC 1456)
1049	Massey-Ferguson MF1150 Dsl
1050	John Deere 4320 Syncro-Range Dsl
1051	Ford 3000 6-Speed (All Purpose) Gaso
1052	Ford 3000 6-Speed (All Purpose) Dsl
1053	Ford 2000 6-Speed (All Purpose) Dsl
1054	Ford 2000 6-Speed (All Purpose) Gaso
1055	Oliver 1955 Dsl
1056	Oliver 1755 Gaso
1057	Oliver 1755 Dsl
1058	Case 770 Power Shift Dsl
1059	Case 870 Power Shift Gaso
1060	Case 870 Manual Gaso
1061	Case 770 Manual Gaso
1062	Case 1170 Dsl
	(Also 1175)

1971

1063	John Deere 7020 Dsl
1064	John Deere 4620 Power-Shift Dsl
1065	Allis-Chalmers Landhandler Two-Ten Dsl
1066	Case 1070 Power-Shift Dsl
	(Also 1090 PS)
1067	Case 1070 Manual Dsl
	(Also 1090 Manual)
1068	Case 970 Power-Shift Dsl
1069	Minneapolis-Moline G-1350 Dsl
	(Also Oliver 2155)
1070	Minneapolis-Moline A4T-1600 Dsl
	(Also Oliver 2655)
1071	Kubota L-210 Dsl
1072	Kubota L-260 Dsl
1073	John Deere 4620 Syncro-Range Dsl
1074	Deutz 8006 Dsl
1075	Deutz 4006 Dsl
1076	Case 870 Power-Shift Dsl
1077	Case 870 Manual Dsl
1078	Case 970 Manual Dsl
1079	Steiger Bearcat 4-WD Dsl
1080	IHC Farmall 1466 Turbo Dsl
1081	IHC Farmall 1066 Turbo Dsl
1082	IHC Farmall 966 Dsl
1083	IHC Farmall 1066 Hydro Turbo Dsl
1084	John Deere 2030 Gaso
1085	John Deere 2030 Dsl
1086	Massey-Ferguson MF1500 4-WD Dsl
1087	Massey-Ferguson MF1800 4-WD Dsl
1088	Case 770 Manual Dsl
1089	Case 770 Power-Shift Dsl
1090	Withdrawn

1972

1091	Deutz D60 06 Dsl
1092	Deutz D130 06 Dsl
1093	Ford 7000 Dsl
	(Also 7000 Row Crop)
1094	IHC Farmall 766 Gaso
1095	IHC Farmall 966 Hydro Dsl
1096	International 454 Gaso
1097	International 454 Dsl
1098	IHC Utility 574 Gaso
	(Also IHC Row Crop)
1099	IHC Utility 574 Dsl
	(Also IHC 574 Row Crop)
1100	John Deere 6030 Dsl
1101	John Deere 7520 4-WD Dsl
1102	Case 1370 Dsl
1103	Case 1270 Dsl
1104	Deutz D55 06 Dsl
1105	Deutz D100 06 Dsl
1106	Satoh S650 Gaso
1107	Long R9500 Special Dsl
1108	Long U445 Dsl
	(Also UTB U445)
1109	Long U550 Dsl
	(Also UTB U550)
1110	John Deere 4430 Quad-Range Dsl
1111	John Deere 4030 Quad-Range Dsl
1112	John Deere 4230 Quad-Range Dsl
1113	John Deere 4630 Power-Shift Dsl
1114	Case 2470 Dsl
1115	International 354 Gaso
1116	International 4166 Dsl
1117	International Farmall 766 Dsl
1118	International Farmall 1468 Dsl

1973

1119	Allis-Chalmers 7030 Dsl
1120	Allis-Chalmers 7050 Dsl
1121	Ford 8600 Dual Power Dsl
	(Also Ford 8600 Dual Power Row Crop, 8600 8-Speed All Purpose, 8600 8-Speed Row Crop)
1122	Ford 9600 Dual Power Dsl
	(Also Ford 9600 Dual Power Row Crop, 9600 8-Speed All Purpose, 9600 8-Speed Row Crop)
1123	International 966 Dsl
1124	International 1066 Turbo Dsl
1125	International 1466 Turbo Dsl
1126	International 464 Gaso
1127	International 464 Dsl
1128	International 674 Utility Gaso
	(Also IHC 674 Row Crop)
1129	International 674 Utility Dsl
	(Also IHC 674 Row Crop)
1130	Deutz D130 06 Dsl
1131	Deutz D45 06 Dsl
1132	Massey-Ferguson 1105 Dsl
1133	Massey-Ferguson 1085 Dsl
1134	Massey-Ferguson 1155 Dsl
1135	Massey-Ferguson 1135 Dsl
1136	David Brown 995 Dsl
1137	David Brown 885 Dsl
1138	David Brown 1212 Dsl
1139	Belarus MTZ 80 Dsl

1140	Oliver 2255 Dsl
1141	Minneapolis-Moline G1355 Dsl
1142	David Brown 1210 Dsl
1143	David Brown 885 Gaso
1144	David Brown 990 Dsl
1145	Kubota L225 Dsl
1146	John Deere 830 Dsl
1147	John Deere 1530 Dsl
1148	Case 1370 Dsl
1149	Case 870 Manual Dsl
1150	Case 870 Power-Shift Dsl
1151	International 666 Dsl
1152	International 666 Gaso
1153	International 4366 Turbo Dsl
1154	International Hydro 70 Gaso
1155	International Hydro 70 Dsl

1974

1156	Allis-Chalmers 175 Gaso
1157	John Deere 2630 Dsl
1158	International Hydro 100 Dsl
1159	White Field Boss 4-150 Dsl
1160	Case 1270 Dsl
1161	Minneapolis-Moline G955 Dsl
1162	Steiger Bearcat II Dsl
1163	Long 900 Dsl
1164	Withdrawn
1165	Case 2670 Dsl
1166	Allis-Chalmers 7040 Dsl
1167	Allis-Chalmers 7060 Dsl
1168	Allis-Chalmers 7080 Dsl
1169	Steiger Tiger II Dsl
1170	Steiger Cougar II Dsl

1975

1171	Massey-Ferguson MF285 Std MP Dsl
1172	Massey-Ferguson 1505 Dsl
1173	Massey-Ferguson 1805 Dsl
1174	International 1566 Turbo Dsl
1175	International 1568 Dsl
1176	Leyland 255 Dsl
1177	Leyland 245 Dsl
1178	Leyland 2100 Dsl
1179	John Deere 8430 Dsl
1180	John Deere 8630 Dsl
1181	White Field Boss 2-105 Dsl
1182	White Field Boss 2-150 Dsl
1183	Belarus 250 Dsl
1184	White Field Boss 4-180 Dsl
1185	SAME Panter 4-WD Dsl
1186	SAME Buffalo 4-WD Dsl
1187	Massey-Ferguson MF235 Dsl
1188	Massey-Ferguson MF265 Dsl
1189	Massey-Ferguson MF255 Dsl
1190	Massey-Ferguson MF255 Gaso
1191	John Deere 2040 Dsl
1192	John Deere 2240 Dsl
1193	Massey-Ferguson MF275 Dsl
1194	Massey-Ferguson MF235 Gaso
1195	Allis-Chalmers 7000 Power-Shift Dsl
1196	Allis-Chalmers 7040 Power-Shift Dsl
1197	Allis-Chalmers 7060 Power-Shift Dsl
1198	Kubota L285 Dsl
1199	Yanmar YM240 Dsl
1200	Belarus T-25A Dsl